ISSUES IN THE IMPACTS OF CLIMATE VARIABILITY AND CHANGE ON AGRICULTURE

Applications to the southeastern United States

ISSUES IN THE IMPACTS OF CLIMATE VARIABILITY AND CHANGE ON AGRICULTURE

Applications to the southeastern United States

Edited by

Linda O. Mearns

National Center for Atmospheric Research,
Boulder, CO, U.S.A.

Reprinted from Climatic Change
Volume 60, Nos. 1–2, 2003

Kluwer Academic Publishers
Dordrecht / Boston / London

A C.I.P. Catalogue record for this book is available from the Library of Congress.

ISBN 1-4020-1618-2

Published by Kluwer Academic Publishers,
P.O. Box 17, 3300 AA Dordrecht, The Netherlands.

Sold and distributed in North, Central and South America
by Kluwer Academic Publishers,
101 Philip Drive, Norwell, MA 02061, U.S.A.

In all other countries, sold and distributed
by Kluwer Academic Publishers,
P.O. Box 322, 3300 AH Dordrecht, The Netherlands.

Cover photographs:
Corn photograph by Warren Gretz courtesy of DOE/NREL, Colorado, U.S.A.
Cotton photograph by Dan Mott
Sorghum photograph courtesy of Max-Planck-Institute for Plant Breeding, Cologne, Germany
Thunder cloud photograph courtesy of NOAA/AOML/Hurricane Research Division, Colorado, U.S.A.

Printed on acid-free paper

Printed in The Netherlands

Contents

ISSUES IN THE IMPACTS OF CLIMATE VARIABILITY AND CHANGE ON AGRICULTURE

Applications to the southeastern United States
Guest Editorial

As the title of this special issue suggests, the commonality of the papers included herein is a focus on the interaction of agriculture with climate variability and change in the southeastern United States. The range of topics covered is considerable, from climate modeling to remote sensing to economics. All papers result from two major projects funded by NASA MTPE and the U.S. EPA NCQERA. In addition, the USDA-ERS funded part of one study concerning economics of agriculture under climate change.

We chose to study the Southeast because it possesses characteristics that we assumed would further our main project interests, which included exploration of the uncertainty of spatial scale of climate scenarios, uncertainties in modeling adaptation in agricultural assessment work, and exploration of relationships between large scale climate modes, vegetation condition, and local daily weather variables. A central question thus arises: do the papers represent true regional studies, being fundamentally about the southeastern U.S., or do the papers concern methodological/conceptual studies, wherein the Southeast should be viewed as an apt application region? Most of the papers are hybrids, embodying characteristics of both types of studies (regional and methodological), but on balance the Southeast serves more as an apposite application region.

The Southeast is an agriculturally diverse region, where a wide variety of crops are grown. It produces about 40% of the national total value of cotton, 23% of rice, and a considerable portion of the total value for most vegetables and citrus fruits. While not significant in terms of total national production, wheat, corn, soybean, and sorghum are also produced (Hansen et al., 2001). It is a region that could be highly vulnerable to climate change given its current climate, which is probably above the optimum (in terms of temperature), for such crops as wheat and soybeans. It is also a region with physiographic characteristics that suggest it might benefit from application of higher resolution scenarios for a climate change assessment of agriculture. Such characteristics include the presence of mountains (Appalachians), complex land-use patterns, and complex coastlines. Furthermore, it is a region that has been considerably studied from the point of view of the influence of large scale climate modes, such as El Niño-Southern Oscillation, on its climate, vegetation, and crop production. This collection of characteristics and previous research indicated thus that the Southeast would act as an excellent application region for our research goals.

The first five papers form an integrated project concerning the uncertainty in agricultural and economic impacts that result from varying scales of climate change information. The issue of spatial scale of climate scenarios has long been a topic in climate impacts analysis (Gates, 1985), because of the mismatch of spatial scale between the climate information available from global climate models (on the order of 100s of kms) and the scale needed by many impacts models (on the order of meters to 10s of kms). In the past ten years or so techniques have been developed to provide higher resolution information regarding climate change, such as regional climate modeling and statistical downscaling (Giorgi and Mearns, 1991, 1999). Yet only recently have studies been pursued that carefully examine the effect of higher resolution information (in contrast to coarse resolution) on calculated agricultural impacts of climate change on various temporal and spatial scales (Mearns et al., 1999, 2001; Easterling et al., 2001, Guereña et al., 2001).

The project presented in these five papers takes these types of studies one step further by integrating crop model results in the context of regional economic impacts of changes in crop yields. We apply climate, crop, and economic models to explore the significance of spatial scale of scenarios for an agricultural integrated assessment. The first paper (Mearns et al., 2003) describes the climate scenarios, the next three papers (Tsvetsinskaya et al., 2003; Carbone et al., 2003; Doherty et al., 2003) present the crop modeling results, and the fifth paper (Adams et al., 2003) the economic results.

Coarse and fine scale climate scenarios for the southeastern U.S. were used, formed from general circulation model (GCM) control and $2 \times CO_2$ experiments and those of a regional climate model that employed boundary conditions from the GCM experiments. Thus, the two scenarios are dynamically related, and exhibited similar climate changes on a very broad regional scale, but detailed subregional changes were sometimes quite different. For example, both scenarios exhibited decreased precipitation in the summer, but these decreases were larger in the high resolution model, especially on the coastal plain.

The two scenarios, with horizontal resolutions of 300 and 50 km, respectively, were applied to a series of crop models representing many of the crops grown in the Southeast (corn, cotton, rice, sorghum, soybean, and wheat) to determine the effect of the resolution of the scenarios on the calculations of changes in yield (from baseline conditions) at various levels of spatial aggregation (all of the Southeast, state level, and 50 km grid level).

Most of these crop types have been modeled in numerous climate change impacts studies. Cotton, however, has less commonly been studied. Our project presents the first region-wide application of a complete cotton model in a climate change study. The changes in crop yields were then used in an agricultural economic model to determine if the spatial scale of the scenarios affects regional economics. For this latter study, crop yield changes for the other major cropping areas of the United States were also calculated, using two other sets of regional climate model runs that covered the rest of the U.S. These runs used the same regional model as for the Southeast and were driven by boundary conditions from the same GCM.

Results indicate that the effect of the scale of climate change scenarios varies with the particular crop considered, as well as the scale of aggregation of the cropping results and the management treatment (with or without adaptation measures). Some simulated crops, such as cotton, soybean, and sorghum, experienced substantial contrasts in changes in yield, regardless of the spatial aggregation level of the yield results. Others, such as corn, showed little contrast when results were aggregated to the entire Southeast region, but exhibited significant contrasts on the state and 50-km level. Winter wheat, however, exhibited very little contrast regardless of the aggregation level. Each cropping paper analyzes, for the individual crops, which differences in the climate changes are most responsible for the differences in the changes in yields. For virtually all crops, adaptation tends to mitigate the contrasting effects of the scenarios, as would be expected, since adaptation attempts to minimize the effect of all adverse climate conditions.

When all cropping results were used as input to the agricultural economic model, ASM, the coarse scale scenario clearly demonstrated a more positive effect on the overall national economic well-being in the agricultural sector, both with and without adaptation. When considering regional productivity, some regions, such as the Southeast, fare considerably worse than others, and one might expect to see agriculture, aside from cotton production, diminish as an economic force in the region as a whole. The northern and southern plains and the Pacific coast states fare better with the high resolution scenario, but the opposite is true for all other regions of the U.S.

The over arching conclusions of this integrated project are that the scale of scenario matters both in terms of changes in crop yields and in final national and regional economic results. Indeed, we were surprised that the level of aggregation and 'filtering' of the crop yields in the economic model did not essentially wash out the contrasting effects of the climate scenarios.

It is important to note that we do not claim that one scenario is in any way more plausible than the other. It must be remembered that in contrast to scenarios constructed from more recent transient climate model runs, our scenarios use equilibrium doubled CO_2 runs, and thus are less complete. However, evidence is mounting that higher resolution scenarios likely provide more realistic responses to changes in forcing (e.g., due to increased greenhouse gases) than coarse scale scenarios in regions with high relief (mountainous areas), complex coastlines, or areas with complex land-use patterns (Giorgi et al., 2001).

The southeastern United States contains elements of all three physiographic conditions. Resolving the Appalachian Mountains with the regional climate model affects the final climate changes simulated. Florida does not exist as a landmass in the coarse scale GCM, and land use in the Southeast is much more spatially complex than in the central Great Plains, for example. Hence this study demonstrates the importance of considering spatial scale of scenarios when constructing climate impact assessments, at least for agriculture, and we strongly recommend that more experiments exploring the effect of scenario spatial scale be performed.

Christensen, J. H., Carter, T. R., and Giorgi, F.: 2002, 'PRUDENCE Employs New Methods to Assess European Climate Change', *EOS Transactions* **83** (13), 147.

Doherty, R. M., Mearns, L. O., Reddy, K. R., Downton, M., and McDaniel, L.: 2003, 'Spatial Scale Effects of Climate Scenarios on Simulated Cotton Production in the Southeastern U.S.A.', *Clim. Change* **60**, 99–129.

Easterling, W. E., Chhetri, N., and Niu, X.: 2003, 'Improving the Realism of Modeling Agronomic Adaptation to Climate Change: Simulating Technological Substitution', *Clim. Change* **60**, 149–173.

Easterling, W., Mearns, L. O., Hays, C., and Marx, D.: 2001, 'Comparison of Agricultural Impacts of Climate Change Calculated from High and Low Resolution Climate Model Scenarios: Part II. The Effects of Adaptation', *Clim. Change* **51**, 173–197.

Gates, W. L.: 1985, 'The Use of General Circulation Models in the Analysis of the Ecosystem Impacts of Climatic Change', *Clim. Change* **7**, 267–284

Giorgi, F. and Mearns, L. O.: 1991, 'Approaches to the Simulation of Regional Climate Change: A Review', *Rev. of Geophysics* **29**, 191–216.

Giorgi, F. and Mearns, L. O.: 1999, 'Regional Climate Modeling Revisited: An Introduction to the Special Issue', *J. Geophys. Res.* **104** (D6), 6335–6352.

Giorgi, F. et al.: 2001, 'Regional Climate Information: Evaluations and Projections', Chapter 10, in Houghton et al. (eds.), *IPCC Third Assessment Report. The Science of Climate Change*, Cambridge University Press, pp. 583–638.

Guereña, A., Ruiz-Ramos, M., Diaz-Ambrona, C., Conde, J., and Minguez, M.: 2001, 'Assessment of Climate Change and Agriculture in Spain Using Climate Models', *Agron. J.* **93**, 237–249.

Hansen, J. W., Jones, J. W., Irmak, A., and Royce, F.: 2001, 'El Niño-Southern Oscillation Impacts on Crop Production in the Southeast United States', in *American Society of Agronomy, Impact of El Niño and Climate Variability on Agriculture*, ASA Special Publication No. 63, pp. 55–76.

Katz, R. W., Parlange, M. B., and Tebaldi, C.: 2003, 'Stochastic Modeling of the Effects of Large-Scale Circulation on Daily Weather in the Southeastern U.S.', *Clim. Change* **60**, 189–216.

Mearns, L. O., Easterling, W., Hays, C., and Marx, D.: 2001a, 'Comparison of Agricultural Impacts of Climate Change Calculated from High and Low Resolution Climate Model Scenarios: Part I. The Uncertainty of Spatial Scale', *Clim. Change* **51**, 131–172.

Mearns, L. O., Giorgi, F., McDaniel, L., and Shields, C.: 2003, 'Climate Scenarios for the Southeastern U.S. Based on GCM and Regional Model Simulations', *Clim. Change* **60**, 7–35.

Mearns, L. O., Hulme, M., Carter, T. R., Leemans, R., Lal, M., and Whetton, P.: 2001b, 'Climate Scenario Development', Chapter 13 in Houghton et al. (eds.), *IPCC Third Assessment Report. The Science of Climate Change*, Cambridge University Press, pp. 739–768.

Mearns L. O., Mavromatis, T., Tsvetsinskaya, E., Hays, C., and Easterling, W. E.: 1999, 'Comparative Responses of EPIC and CERES Crop Models to High and Low Spatial Resolution Climate Change Scenarios', *J. Geophys. Res.* **104**, 6623–6646.

Peters, A. J., Ji, L., and Walter-Shea, E.: 2003, 'Southeastern U.S. Vegetation Response to ENSO Events (1989–1999)', *Clim. Change* **60**, 175–188.

Tsvetsinskaya, E. A., Mearns, L. O., Mavromatis, T., Gao, W., McDaniel, L. R., and Downton, M. W.: 2003, 'The Effect of Spatial Scale of Climatic Change Scenarios on Simulated Maize, Winter Wheat, and Rice Production in the Southeastern United States', *Clim. Change* **60**, 37–71.

National Center for Atmospheric Research,
Environmental and Societal Impacts Group,
P.O. Box 3000,
Boulder, CO 80307-3000,
U.S.A.

LINDA O. MEARNS
Guest Editor

CLIMATE SCENARIOS FOR THE SOUTHEASTERN U.S. BASED ON GCM AND REGIONAL MODEL SIMULATIONS

L. O. MEARNS [1], F. GIORGI [2], L. McDANIEL [1] and C. SHIELDS [1]

[1] *National Center for Atmospheric Research,⋆ Boulder, Colorado, U.S.A.*
[2] *Abdus Salam International Centre for Theoretical Physics, Trieste, Italy*

Abstract. We analyze the control runs and 2 × CO_2 projections (5-year lengths) of the CSIRO Mk 2 GCM and the RegCM2 regional climate model, which was nested in the CSIRO GCM, over the Southeastern U.S.; and we present the development of climate scenarios for use in an integrated assessment of agriculture. The RegCM exhibits smaller biases in both maximum and minimum temperature compared to the CSIRO. Domain average precipitation biases are generally negative and relatively small in winter, spring, and fall, but both models produce large positive biases in summer, that of the RegCM being the larger. Spatial pattern correlations of the model control runs and observations show that the RegCM reproduces better than the CSIRO the spatial patterns of precipitation, minimum and maximum temperature in all seasons. Under climate change conditions, the most salient feature from the point of view of scenarios for agriculture is the large decreases in summer precipitation, about 20% in the CSIRO and 30% in the RegCM. Increases in spring precipitation are found in both models, about 35% in the CSIRO and 25% in the RegCM. Precipitation decreases of about 20% dominate in winter in the CSIRO, while a more complex pattern of increases and decreases is exhibited by the regional model. Temperature increases by 3 to 5 °C in the CSIRO, the higher values dominating in winter and spring. In the RegCM, temperature increases are much more spatially and temporally variable, ranging from 1 to 7 °C across all months and grids. In summer large increases (up to 7 °C) in maximum temperature are found in the northeastern part of the domain where maximum drying occurs.

1. Introduction

Climate change experiments with regional climate models (RCMs) nested in coarser resolution general circulation models (GCMs) have become common particularly over the past half decade (Giorgi et al., 2001), and are now being used to form climate scenarios for input to impact models (Mearns et al., 2001). This development has been welcomed by the climate impacts community since dissatisfaction with the coarse spatial scale of general circulation model (GCM) simulations used to form climate scenarios for impacts use has been widely expressed (Gates, 1985; Cohen, 1990; Carter et al., 1994). It is perceived that there is a serious mismatch in scale between that of the global climate models (100s of km) and the scale of concern for most regional impact assessments (at least an order of magnitude finer). For example, crop models operate at spatial resolutions of a single plant to a

⋆ The National Center for Atmospheric Research is sponsored by the National Science Foundation.

Climatic Change **60:** 7–35, 2003.

few meters to hectares. Fine scale variations in climate not simulated by GCMs are potentially important to the scale of concern for various resources such as agriculture.

In this paper we present climate model simulations that were expressly created to provide climate scenarios for a study of the effect of spatial scale of climate scenarios on agriculture in the southeastern United States, the subject of many of the papers in this Special Issue of Climatic Change. We describe and analyze the coarse (from the CSIRO GCM) and fine scale (from the regional model RegCM) climate simulations, and present the formation of the scenarios to act as input to a number of crop models for an intergrated assessment of climate impacts on agriculture in the southeastern U.S.

The analysis presented in this paper focuses on precipitation and near-surface air temperature, usually the two most important variables used in crop models. We analyze the control runs of both the GCM and the regional climate model (RCM) regarding model biases and provide some dynamical explanations for the control runs' departures from observed climatology. We then go on to discuss the doubled CO_2 simulations, particularly the simulated changes in precipitation and temperature that form the basis of the climate scenarios created for use in the crop models. We stress that the simulations are not intended to provide 'predictions' of climate change. This would require ensembles of transient simulations with different GCMs and nested regional models. Rather they provide the perturbed climate input for use in an uncertainty analysis of the effect of spatial scale of climate scenarios on an integrated assessment of agriculture.

In Section 2 we describe the climate models and the climate runs produced. The control runs are analyzed and validated in Section 3, while the doubled CO_2 runs and scenarios formation are discussed in Section 4. In Section 5 we consider signal-to-noise issues related to the run lengths. Summary and conclusions are presented in Section 6.

2. Models and Experiments

2.1. DESCRIPTION OF MODELS USED

For this work we used the NCAR RegCM of Giorgi et al (1993a,b) driven by output from the control and equilibrium $2 \times CO_2$ simulations with the CSIRO Mark 2 GCM (Watterson, 1998; Watterson et al., 1997, 1999). The GCM was run at R21 horizontal spectral resolution (about 5° in physical space) with nine levels in the vertical and complete physics representation. It uses an Arakawa moist adjustment parameterization of deep convection and includes shallow convection. Cloud cover is determined in three layers and is a function of relative humidity. In addition to parameterizations for the surface boundary layer, soil moisture, and snow cover, it includes ice dynamics with a prognostic open-water fraction in sea ice (Watterson et al., 1997). Land grid squares are partitioned into bare soil and vegetation fractions (Kowalczyk et al., 1994). The atmospheric model is coupled to a 50 m depth mixed layer ocean. Thirty years of control (using a CO_2 concentration of 330 ppmv) and doubled CO_2 runs were produced. The global mean surface temperature increase under $2 \times CO_2$ conditions is 4.3 °C and global precipitation increases by 10%. The global climatology of the control and $2 \times CO_2$ runs are described in Watterson (1998) and Watterson et al. (1999).

The RegCM is an augmented version of the NCAR/Pennsylvania State University mesoscale model MM4 (Giorgi et al., 1993a,b). It is a primitive equation σ vertical coordinate, grid point limited-area model with compressibility and hydrostatic balance. Physics parameterizations incorporated into the model for application to climate studies include the BATS surface package (Biosphere-Atmosphere Transfer Scheme, Dickinson et al., 1993), an explicit planetary boundary layer formulation (Holtslag et al., 1990), a detailed atmospheric radiative calculation package (Briegleb, 1992), a mass flux cumulus parameterization scheme (Grell et al., 1994), and a simplified explicit moisture scheme including an equation for cloud water (Giorgi and Marinucci, 1996). The RegCM was previously nested in the same CSIRO experiments over the western two-thirds of the U.S. (Giorgi et al., 1998).

The RegCM model domain covers the southeastern U.S. using a Lambert conformal projection with a horizontal grid point spacing of 50 km (Figure 1). Fourteen vertical levels are used between the model top (80 mb), with a vertical resolution of $\sigma = 0.1$ in the troposphere and 5 levels below 1500 meters. The closest sigma level to the 1500 m height is 0.815. The land-use distribution for the model domain was derived from a U.S. Geological Survey data set obtained from a combination of remote sensed and ground based data (Loveland et al., 1991).

The high spatial resolution of the model allows for a clear representation of the Appalachian Mountains and the peninsula of Florida, which is not possible in the GCM (Figures 1 and 2). There is no land grid point in the GCM corresponding to Florida.

2.2. RegCM Runs

Two continuous simulations were carried out, a 5-year present day (control) run and a five year $2 \times CO_2$ with the RegCM driven by time-dependent lateral meteorological fields from the corresponding CSIRO simulations. The same CO_2 concentrations used in the CSIRO were used in the RegCM runs (330 ppmv and 660 ppmv, respectively). Although longer simulations are preferable to increase the statistical significance of the climate signals, previous experience has shown that the interdecadal variability of equilibrium runs is lower than for transient runs, so that even relatively short simulation periods are sufficient to represent the basic behavior of the models (Giorgi et al., 1994, 1998).

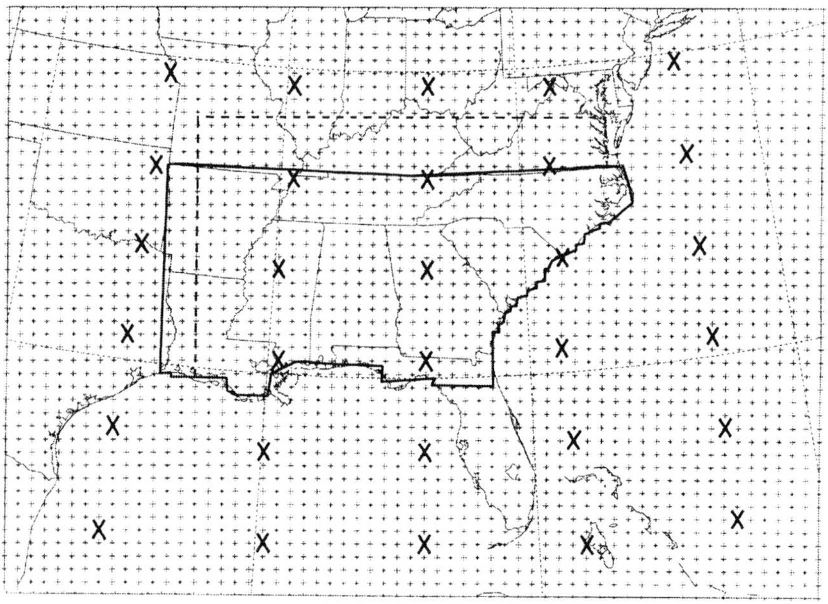

Figure 1. The domain of the regional climate model (RegCM) experiments showing grid points of the RegCM (crosses) and grid points of the CSIRO GCM (large Xs). The area outlined with a heavy solid line is the actual area of the crop study, and the dashed line indicates the climate validation area.

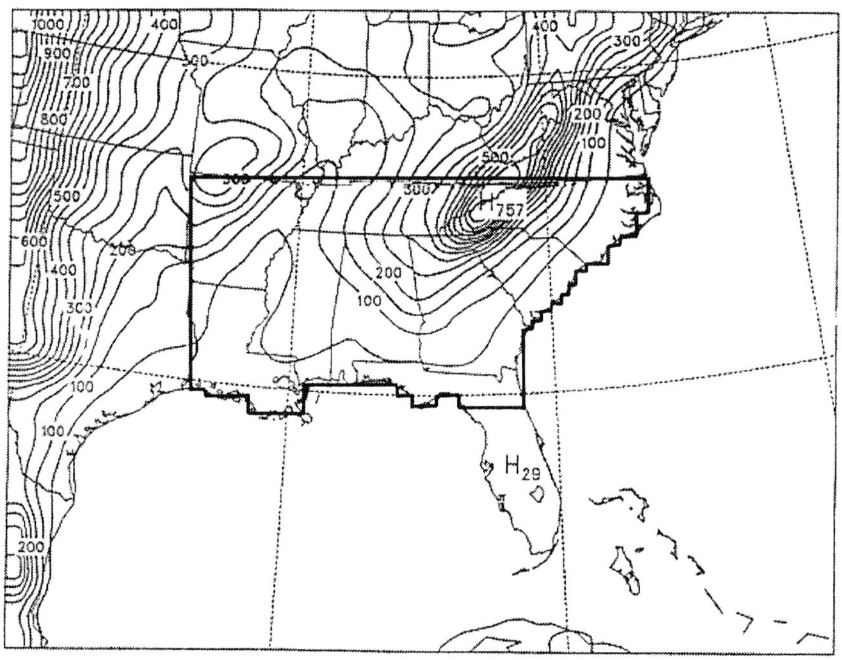

Figure 2. The topography of the RegCM. Contour interval is 50 meters. The area outlined with a heavy solid line is the actual area of the crop study.

The nesting technique is a standard relaxation method for wind, temperature, water vapor and surface pressure applied over a lateral buffer area of 10 grid points, and uses an altitude dependent exponential weighting coefficient for the relaxation terms (Giorgi et al., 1993b). The GCM forcing data were provided at 8 hour intervals, with linear interpolation to each model time step (3 minutes). In the RegCM runs, initial soil water content was interpolated from the GCM normalized soil water content, and time dependent sea surface temperatures were interpolated from the ocean component of the GCM.

Although we do not have available a RegCM simulation driven by analyses of observations for the present domain, an indication of the model performance over the eastern U.S. can be obtained from the work of Giorgi and Shields (1999). They carried out a three year simulation for a domain encompassing the whole continental U.S. with the RegCM driven by ECMWF analyses of observations. Giorgi and Shields (1999) present an analysis of this experiment for different subregions, one of which is the Eastern U.S. They show that the model exhibits seasonally averaged biases in the range of –25.3% (winter) to 17.9% (spring) for precipitation and –1.66 °C (summer) to –0.66 °C (winter) for surface air temperature. Therefore, the model exhibits a generally good simulation of seasonal precipitation and a tendency of a cold bias of a few degrees or less over the region. Giorgi and Shields (1999) also show that the model driven by analyses of observations performs well in reproducing the interannual variability of both seasonal precipitation and temperature over the Eastern U.S.

2.3. DEVELOPMENT OF THE OBSERVED DAILY CLIMATE DATA

We developed the data set to serve as a validation data set for the regional climate model in the Southeast, and to serve as the climatological input for crop models that were to be applied over the Southeast under observed and perturbed climate conditions. We developed the observed data set on the grid of the RegCM, covering the states that were to be included in the impacts study (outlined on Figure 1). We did not include the lower peninsula of Florida since the density of stations was low, and many sites had large amounts of missing data. Moreover, this part of Florida, where mainly vegetables are grown, was not highly relevant to the crops we planned to model. Furthermore, since there is no comparable land area in the CSIRO GCM, use of this area in the comparative validations of the control runs would lead to spurious results, and it would be difficult to develop a credible climate change scenario from the CSIRO for this area, which is represented by an ocean grid point. Variables on a daily time scale included maximum and minimum temperature, precipitation, relative humidity, solar radiation, and wind speed. Temperature and precipitation were taken from the U.S. daily cooperative station data set available at NCAR (DS 510.0), which corresponds to the NCDC data set TD3200 U.S. Control Summary of Day. Only stations having less than 10% missing precipitation data for the time period 1960–1995 were considered for use. The station having less than 10% missing data and closest to the center of each RegCM 50 km grid was selected to represent the grid. Missing precipitation data were filled by taking the median of precipitation for stations (three or more) within a 0.5 degree radius from the primary station. Missing temperature data were filled by taking the average of temperature from these surrounding stations. The other variables, solar radiation, relative humidity, and wind speed were generated using the weather generator available with one of the crop models used (Richardson and Nicks, 1990). Mean monthly data for these variables available as part of an EPIC crop model data set also served as input to the weather generator. Generated solar radiation data were spot-checked using the sparse network (30 stations) of daily observed radiation stations in the Southeast (NREL, 1992).

3. Evaluation of Control Runs

We evaluated the control runs of the RegCM and CSIRO using several different observed data sets: the daily data set (described in Section 2.3 above), the Legates and Willmott data sets (Legates and Willmott, 1990a,b, referred to as LW) and the University of East Anglia (UEA) data set (New et al., 1999, 2000). The LW data set was used initially while running the regional climate model to evaluate the model performance for each year of the run. That data set includes only mean monthly temperatures and precipitation, assembled from data sets covering varying time periods, but includes both land and ocean. The daily data set is used with a statistics package (Mearns et al. 1995a,b) to statistically evaluate both the CSIRO and RegCM control runs and other contrasts. Since the model runs are relatively short (5 years), it is particularly desirable to use statistics that consider daily data.

We used the University of East Anglia (UEA) (New et al., 1999, 2000) data set to expand the coverage of the evaluation, since the daily data set was only developed for the part of the regional climate domain that was used for the impacts study. The UEA data set, which is a monthly data set, developed on a 0.5 degree grid, covers 1900–1995. We used the subset of years 1960–1995, the same set of years as the daily data set. The UEA data set contains monthly time series of mean maximum and minimum temperature and precipitation for the full time series. However, it is a land only data set. The UEA monthly data set was bilinearly interpolated to the RegCM grid, which has a similar spatial resolution.

The arrangement of the CSIRO grids is such that only 5 complete grids are contained in the impacts study area. Using the UEA data we were able to look at a larger portion of the domain, covering 8 complete CSIRO grids to get a more complete picture of the CSIRO model performance (outline on Figure 1). We did not include the western most CSIRO grids in our evaluation of control runs since only their eastern edges were in the study area, and the far western parts of these grids fell within the interpolation buffer zone. We compared the daily data set and the UEA data set on a monthly and seasonal time scale and found them to be very similar both for temperature and precipitation. Precipitation estimates in general differed by 1% or less and temperatures by 0.25 °C.

3.1. DOMAIN WIDE EVALUATION USING THE UEA DATA SET

The CSIRO was originally chosen for this nesting work since it reproduced relatively well the general climatology over the continental U.S. as represented by the L&W data set.

Reproduction of the climatology of the southeastern U.S. is often challenging for climate models because of a number of factors: the importance of convective precipitation in the region; the complexity of the moisture sources for precipitation (Gulf of Mexico and the Atlantic Ocean); the location and strength of the Bermuda high; the location and strength of the nocturnal jet; and the importance of contributions to precipitation from hurricanes along the coast in late summer (Robinson and Henderson, 1992; Henderson and Vega, 1996).

As a result of these various processes, precipitation over the southeastern U.S. shows different seasonality over different areas of the region (see Figures 3a–d). In the winter (Figure 3a), precipitation is maximum over Louisiana and the southern half of Mississippi and Alabama. Maximum precipitation is also found over the Appalachians, with precipitation decreasing towards the north and west of the region. The basic features of this precipitation pattern do not change much in the spring (Figure 3b), although the westward gradient of precipitation is less pronounced than in the winter. The precipitation patterns change dramatically in the summer (Figure 3c), when maximum precipitation is found over the coastal regions, both on the Gulf of Mexico and the Atlantic. A secondary, but not pronounced maximum in summer precipitation still occurs over the Appalachians, while again precipitation decreases towards the west. Finally, another major shift in precipitation patterns occurs in the fall (Figure 3d), when the region overall experiences its driest conditions, especially over Georgia, South Carolina and North Carolina, and a maximum is shifted towards the west over western Louisiana and Arkansas.

Concerning temperature (not shown), as expected the region shows a north to south positive gradient and a pronounced seasonal cycle. Observed average temperatures over the validation domain varied from about –2 °C to 14 °C in the winter to about 18 °C to 28 °C in the summer. The diurnal temperature range is also pronounced, in the range of 10–14 °C in all seasons. The Appalachians are also clearly identified by the lower temperatures (2–3 °C) compared to the surrounding regions at the same latitudes.

Table I presents the average seasonal biases (using the UEA data as the observed data) for both model control runs over the central domain, which extends further to the north than the study area in order to include all of the northern CSIRO grid boxes.

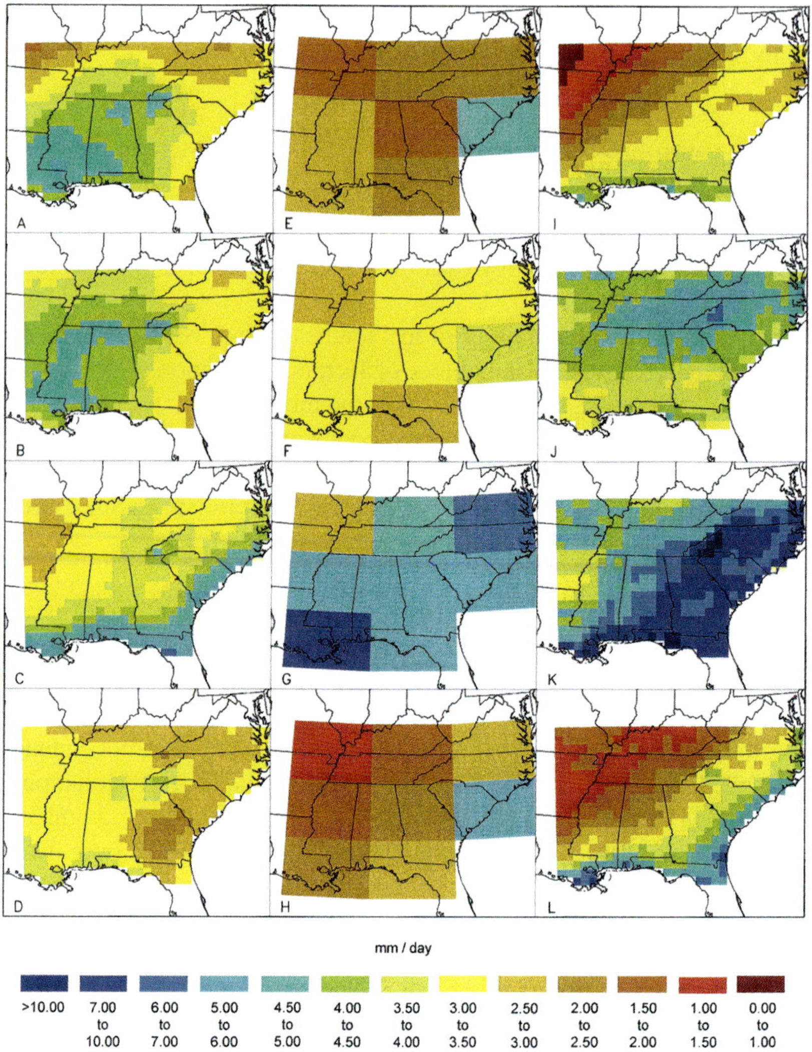

Figure 3. Maps of seasonal precipitation (mm/day) for winter, spring, summer and fall for UEA observations (a–d); CSIRO control run (e–h); and RegCM control run (i–l).

The biases are calculated for each season by averaging over this domain slightly larger than the study area. For temperature we focus on daily maximum and minimum temperature since the average can be affected by compensating errors in the maxima and minima. Overall there are, as is often the case, general similarities in the biases produced by the GCM and the RCM aggregated over the region (Table I). Both models have cold biases in maximum temperature and cold or warm biases in minimum temperature during most of the year. The CSIRO, however, has uniformly larger (and positive) biases in minimum temperature. Note that in all seasons, the biases in the RegCM are smaller than those in the CSIRO, especially

Table I

Southeast domain average biases in the control run climates of the CSIRO and RegCM2 (5 years each), compared with the UEA observed data (1961–1995)

	Maximum temperature (°C)	Minimum temperature (°C)	Precipitation (%)
Winter			
CSIRO	–2.8	0.9	–30.7
RegCM	–2.3	–0.4	–26.4
Spring			
CSIRO	–3.5	1.5	–13.6
RegCM	–2.8	–0.7	16.5
Summer			
CSIRO	–2.4	2.7	36.0
RegCM	–1.6	–0.0	60.6
Fall			
CSIRO	–2.4	2.9	–19.6
RegCM	–1.6	0.8	–8.9

CSIRO indicates results where the UEA data were aggregated to the CSIRO and then the comparison done at that scale.

for the minimum temperature. This improvement is likely not due to the increase in resolution in the RegCM, but rather to the effect of the different land surface schemes in the models. The BATS scheme used in the RegCM and the land surface scheme used in the CSIRO (Kowalczyk et al., 1994) are of similar structure and complexity. However they use different vegetation parameters and drag coefficient formulations, so that these can affect the model results.

Note that Table I implies that the diurnal temperature range as calculated by the RegCM is generally more pronounced than that calculated by the CSIRO. In addition to the use of different land surface schemes, another contribution to this result may come from differences in the simulation of cloudiness.

The spatial patterns of biases in temperature for the RegCM are the following (not shown for brevity). In winter and spring there is a gradient of negative bias in maximum temperature going from values of around –1 °C in the western part of the subregion (western Arkansas) to values up to –5 °C in a small portion of the the northeastern part of the domain (N. Carolina). In summer and fall biases progress from slightly positive in the western part to about –5 °C in Georgia. A possible factor that would contribute to the general underestimation of maximum temperature by the RegCM is related to the simulation of clouds. It has been noted (e.g., Giorgi et al., 1999) that the RegCM has a tendency to simulate excessively high cloud optical thicknesses when clouds occur. Especially in the presence of daytime convective cloudiness this would lead to a decrease in maximum temperatures. Minimum temperatures are generally overestimated in the fall up to 2 °C in the western part of the domain. In summer, minimum temperature biases vary between + and –1 °C, with no set spatial pattern. In spring and winter largest minimum temperature biases are again found in the northeastern part of the domain (N. Carolina) but values do not exceed 3 °C.

Biases in maximum temperature for the CSIRO are negative throughout the southeast area in spring, summer, and fall (not shown). In spring the largest bias is in the northeastern part of the domain (about –4 °C). A similar pattern is seen in the summer. In fall the largest biases are found in the southcentral area (e.g., Alabama). In winter the largest negative biases, around –5 °C, are seen in the northcentral area (Tennessee). The patterns of biases in minimum temperature are similar for spring, summer, and fall, with largest positive biases in the eastern grids, reaching about 4 °C in summer. In winter the biases are mixed positive and negative.

Moving to precipitation (Table I), region-averaged precipitation biases of both models are negative in the winter and fall, and positive in the summer. In the spring the biases are of opposite sign and are relatively small. The biases in the models are similar in the two cold seasons, but the positive bias of the RegCM in summer is larger than that of the GCM. The magnitude of these biases is generally similar to that of the CSIRO and the RegCM for the central Great Plains analyzed by Giorgi et al. (1998), except in the summer for the RegCM.

To better understand these biases we can compare the CSIRO and RegCM seasonal precipitation, shown in Figures 3e–l, with the observations of Figures 3a–d. In winter, both models capture the northwestward negative precipitation gradient towards the Central Plains. However, this extends too far south and east in the models. In addition, precipitation is also undersimulated over the central and eastern regions of our study area, where however, the RegCM results are closer to observations than the CSIRO's.

In the spring the CSIRO underestimates precipitation throughout the domain except for one grid in the east (Figure 3f). The magnitude of RegCM precipitation is more in line with observations (Figure 3j), although the maximum in the RegCM shifts to the northeast of the observed one. Note that the maximum precipitation along the crest of the Appalachians is captured by the RegCM. This is due to orographic uplift of air within the eastward travelling storm systems and possibly to convective activity induced by the solar heating of high elevation surfaces (e.g., Giorgi, 1991).

Differences in warm season precipitation between the RegCM and CSIRO may be due to two reasons; either the use of different convection schemes or the increase in resolution. For example, Giorgi and Marinucci (1996) showed that the same convection scheme tends to produce higher localized precipitation events as the resolution increases. These events are to some extent self-feeding through the release of condensation heat. This process may be responsible for the greater precipitation amounts found in the RegCM over the Appalachian region in addition to the orographic uplift and high elevation surface heating effects. A possible mechanism for the northeastward shift of the precipitation maximum, which has been also noted in previous RegCM runs (e.g., Giorgi and Shields, 1999) is that the RegCM version used in the present run exhibits deficiences in simulating the effects of mesoscale convective complexes. This problem has been noted in many regional models (e.g., Takle et al., 1999), although parameterizations are now available in the RegCM that considerably ameliorate it (Pal et al., 2000).

Moving now to the summer case, we can notice that, consistent with the observations, the models shift the precipitation maximum to the southeast portions of the study area. It is evident, however, that both models heavily overestimate precipitation over this region. The similarity of the RegCM and CSIRO patterns indicate that the regional model is likely inheriting this bias from the CSIRO and amplifying it possibly through the resolution effect discussed above. Since most of the water vapor input to the region during the summer originates from the Gulf of Mexico, it is likely that the precipitation overestimation is due to excessive southerly water vapor flux.

Finally, both models generally underestimate precipitation during the fall (Figures 3h,l), particularly in the western half of the domain. Again, this seems mostly a result of the bias in the CSIRO simulation which is transmitted to the RegCM. The RegCM produces somewhat higher precipitation amounts over the region than the CSIRO, which brings it closer to observations.

Pan et al. (2001) used essentially the same version of the RegCM as in our experiments and found that the RegCM tended to underestimate cold season precipitation over the southeastern US when driven by NCEP reanalyses and to overestimate it when driven by the HadCM. This illustrates the strong impact of the boundary forcing on the precipitation simulation. In the warm season the precipitation biases were small for the reanalysis-driven runs and positive in the HADCM-driven runs. In all seasons the RegCM tended to simulate greater precipitation amounts than the HADCM, a result in line with the present experiment.

3.2. STATISTICAL TESTS USING DAILY DATA

To determine the statistical significance of the differences described above we used the observed daily data set for the same time period of the smaller domain area and applied a domain statistical package (Mearns et al., 1995a,b) that uses daily data as input. By using statistics for daily data, we take advantage of a a larger sample size than if we used tests for mean monthly data. We analyzed the results of tests on mean daily maximum and minimum temperature and mean daily precipitation for each month. As presented in Katz (1982) statistical inferences regarding mean daily temperature may be made using the following test statistic:

$$Z = \frac{\overline{X}_2 - \overline{X}_1}{[n_2^{-1}\hat{V}^2(2) + n_1^{-1}\hat{V}^2(1)]^{\frac{1}{2}}}, \tag{1}$$

where $\overline{X}_1$ and $\overline{X}_2$ = the estimated means of time series 1 and 2; n_1 and n_2 = sample size of time series 1 and 2; $\hat{V}^2$ = estimated variance of time series 1 or 2 defined by:

$$\hat{V}^2 = \frac{\hat{\sigma}_a^2(p)}{[\sum_{k=0}^{p} \hat{\phi}_k(p)]^2}, \tag{2}$$

where $\hat{\sigma}_a^2(p)$ = estimated innovation variance from an autoregressive process of order p; $\hat{\phi}_k(p)$ = the estimated autocorrelation coefficient of order $p, k = 0, p$.

In this instance we fit a first order autoregressive process to the temperature time series, i.e., $p = 1$. Under the null hypothesis the statistic converges to a standard Gaussian distribution as the sample sizes both tend to infinity.

For precipitation the test statistic is formulated using parameters of both the frequency and intensity aspects of daily mean precipitation. First the right skewed precipitation intensities are logarithmically transformed. The test is constructed with the transformed data.

$$Z = \frac{\overline{Y}_2 - \overline{Y}_1}{[n_2^{-1}\hat{V}^2(2) + n_1^{-1}\hat{V}^2(1)]^{\frac{1}{2}}}, \tag{3}$$

where $\overline{Y}_1$ and $\overline{Y}_2$ = the estimated means of the (log transformed) daily precipitation time series 1 and 2; n_1 and n_2 = sample size of time series 1 and 2; $\hat{V}^2$ = estimated variance of time series 1 or 2.

The estimated variance term is a complex function of transition probabilities (concerning the frequency process) and characteristics of the intensity of precipitation (rain falling on rain days) and is given in Katz (1983) in Equations 31–37.

We use these tests as a general diagnostic to compare the control versus observed climatologies for the two models. There are some statistical issues to consider. First, the tests are performed simultaneously at multiple grids, creating the problem of multiplicity. For any given significance level (e.g., 0.05) one would expect that for at least some of the locations (5%), the null hypothesis would be rejected by chance. Furthermore, there is the problem of spatial correlation of the variables being tested across the grids. We, however, do not use these tests for formal hypothesis testing, but rather as a general diagnostic for comparing the CSIRO and RegCM results. We display the percentage of the domain where significance at the 0.05 or higher level is attained for each model, for each month (Table IIa).

The months that resulted in the largest percentage of the domain with significant (at the 0.05 level) differences between the model control run and observations

Table II
Percent area for precipitation

		Month											
		1	2	3	4	5	6	7	8	9	10	11	12
a. Control versus observations (daily observed dataset)													
CSIRO	+	12.5	12.5	0.0	37.5	75.0	50.0	87.5	87.5	62.5	25.5	12.5	0.0
	–	62.5	50.0	62.5	0.0	0.0	0.0	0.0	0.0	12.5	37.5	50.0	87.5
RegCM	+	0.0	14.1	2.7	48.9	31.3	48.4	88.6	52.1	12.8	3.7	11.9	1.2
	–	56.5	18.0	8.1	0.0	0.2	0.0	0.0	0.0	49.7	56.0	36.8	70.4
b. 2 × CO_2 versus control													
CSIRO	+	0.0	0.0	37.5	62.5	0.0	0.0	0.0	0.0	0.0	27.3	0.0	0.0
	–	25.0	12.5	0.0	0.0	0.0	0.0	37.5	25.0	50.0	0.0	25.0	12.5
RegCM	+	0.5	0.0	3.1	20.7	24.7	4.0	1.1	0.0	0.2	31.6	0.0	4.0
	–	0.0	40.0	0.0	0.0	0.0	8.7	66.5	48.5	32.9	0.2	6.3	0.7

Percentage area across the southeast small domain (covering 8 CSIRO grids) where significant (at the 0.05 level) differences between the two precipitation datasets indicated were found. + and – refer to whether the contrast is positive or negative, i.e., + = significant positive difference and – = significant negative difference. The total % area at the 0.05 level is determined by adding the values (+ and – categories).

for precipitation (RegCM) were July when virtually all of the domain has significant overestimations of precipitation (Table IIa). In the cold season months of September through January, about one half of the domain showed significant underestimations. February and March produced the best results when few grids reported significant differences between the observed and modelled precipitation.

Results for the CSIRO are somewhat more limited because not all relevant grids are completely covered by the daily observed data set. The percentage biases are based on the large CSIRO grids and the aggregation of the daily values to the CSIRO grid scale. Each CSIRO grid represents 12.5% of the area. Months with largest percentages of the area showing overestimations for precipitation include May through September with highest values in July and August. The seasonal pattern is not dissimilar from that of the RegCM, with large underestimations occurring from November through March (Table IIa). March is the month for the RegCM when very good reproduction of the observed precipitation is seen, but essentially there is no corresponding good month for the CSIRO.

It is important to note from the comparison of Table I and Table IIa that, although the region-average precipitation biases in the RegCM are of similar magnitude as those of the CSIRO, the percentage area in which the bias is statistically significant is smaller in the RegCM than in the CSIRO for most months. Even in July, when the bias is much larger for the RegCM, the percentage area in which this bias is significant is essentially the same in the two models. This result indicates that the RegCM reproduces better the spatial distribution of precipitation and that the region-average RegCM bias is more dominated by the contribution of relatively small areas. This is also evident from the comparison of the seasonal spatial patterns of precipitation discussed in the previous section. In the next section we will provide a more quantitative evaluation of the ability of the models to reproduce the observed spatial precipitation patterns.

Regarding temperature, the RegCM minimum temperature best reproduces observations based on the statistical test, when usually more than half of the domain shows no statistically significant difference with observations (not shown). Maximum temperature, however, is significantly underestimated across most of the domain in every month. The CSIRO displays a similar pattern for maximum temperature, which is usually significantly underestimated, particularly November through June. However, minimum temperature is significantly overestimated at all grids from April through October and at most grids for the rest of the months. These results underscore the better reproduction of temperature overall for the RegCM, particularly of minimum temperature, compared to the CSIRO.

3.3. SPATIAL CORRELATIONS

Table III displays the spatial correlations between simulated and observed temperature and precipitation over the domain both for the CSIRO and RegCM. The CSIRO grid results were interpolated to the RegCM grid for this comparison. An adiabatic adjustment for elevation with reference to the RegCM topography is made for the CSIRO temperatures as part of the interpolation procedure.

For the RegCM, the correlations are generally high, greater than 0.80 both for daily maximum and minimum temperature, and the correlation coefficients are comparable to those obtained for the Great Plains in the simulation of the western two-thirds of the United States (Giorgi et al., 1998). RegCM correlations for minimum temperature are better than those for CSIRO, especially in winter when the CSIRO correlation is only 0.66. For maximum temperature, the correlations are comparable in the two models except for summer, when the CSIRO shows a relatively low correlation (0.50) with observations. Overall, Table III shows that the RegCM improves the simulation of the temperature spatial distribution compared to the CSIRO, even though this domain is not characterized by a strong topographical forcing. However, the better representation of the Appalachian mountains and the complex coastlines does contribute somewhat to the improved simulation of temperature. The adiabatic correction of temperature for the CSIRO mitigates the difference in the resolution effect between the two models for the correlations with temperature observations.

RegCM correlations for precipitation are highly variable, ranging from good correlations in summer and to a lesser extent in winter to poor correlations in spring and fall (Table III). These precipitation correlations reflect a poorer spatial repre-

Table III

Spatial correlations observed (UEA) versus CSIRO and RegCM 5-year control runs

	Maximum temperature (°C)	Minimum temperature (°C)	Mean temperature (°C)	Precipitation (%)
Winter				
CSIRO	0.97	0.66	0.93	0.10
RegCM	0.94	0.95	0.95	0.43
Spring				
CSIRO	0.91	0.92	0.95	–0.13
RegCM	0.95	0.96	0.97	0.16
Summer				
CSIRO	0.50	0.85	0.82	0.52
RegCM	0.81	0.96	0.93	0.76
Fall				
CSIRO	0.95	0.86	0.93	–0.21
RegCM	0.86	0.95	0.93	0.01

sentation than was produced for the Great Plains in the aforementioned run (Giorgi et al., 1998), except in summer when results are similar. The spatial correlations are not aligned with the mean biases. The large mean bias in summer, for example, corresponds with a relatively good spatial correlation, whereas the small mean bias in spring corresponds to a poor spatial correlation. This is because, as discussed above, the region-averaged bias may be dominated by the contribution of relatively small regions.

The precipitation correlations of the CSIRO with observations of precipitation are uniformly lower than those for the RegCM in all seasons. However, both models poorly reproduce the pattern of precipitation in spring and fall when, evidently the RegCM cannot compensate for the errors deriving from the large scale forcing. We can conclude from these results that the RegCM overall produces better spatial patterns of temperature and precipitation than does the CSIRO, and in that regard improves on the regional simulation of climate compared to the CSIRO. This result has been found consistently in nested regional climate simulations with different models (Giorgi and Mearns, 1999). However, when the large scale forcing is substantially biased, such as for spring and fall precipitation, the bias is inevitably transmitted also to the regional model.

Table IV

Southeast domain average seasonal climate changes ($2 \times CO_2$ versus control) of the CSIRO and RegCM (5 years each)

	Maximum temperature (°C)	Minimum temperature (°C)	Mean temperature (°C)	Precipitation (%)
Winter				
CSIRO	4.0	4.5	4.3	–18.9
RegCM	3.7	3.7	3.7	–1.5
Spring				
CSIRO	4.6	5.6	5.1	35.7
RegCM	4.1	5.3	4.7	26.5
Summer				
CSIRO	4.2	3.8	4.0	–17.0
RegCM	5.1	3.7	4.4	–30.9
Fall				
CSIRO	4.7	3.9	4.3	–7.2
RegCM	5.2	4.0	4.5	2.7

4. $2 \times CO_2$ Control Climates

4.1. SEASONAL RESULTS

Table IV shows the area-average seasonal temperature and precipitation changes for the CSIRO and RegCM. On an area wide basis most striking changes occur in spring and summer for precipitation, with similar general tendencies in the two models.

Substantial precipitation increases are seen for both models in the spring, but substantial decreases in the summer. The CSIRO experiences larger increases in the spring than the RegCM but the latter model experiences larger decreases in the summer. The precipitation changes in the fall and winter seasons are relatively small, especially in the RegCM. Area wide average changes in temperature are similar in the two models (Table IV). Temperature changes are largest for spring minimum temperature (greater than 5 °C), and smallest for summer minimum temperature.

Detailed spatial contrasts between the climate changes are, however striking. The spatial distribution of the regional changes in precipitation is shown in Figure 4. In spring, the percentage changes in precipitation across the domain are not dissimilar for the two models, both showing increases throughout the area

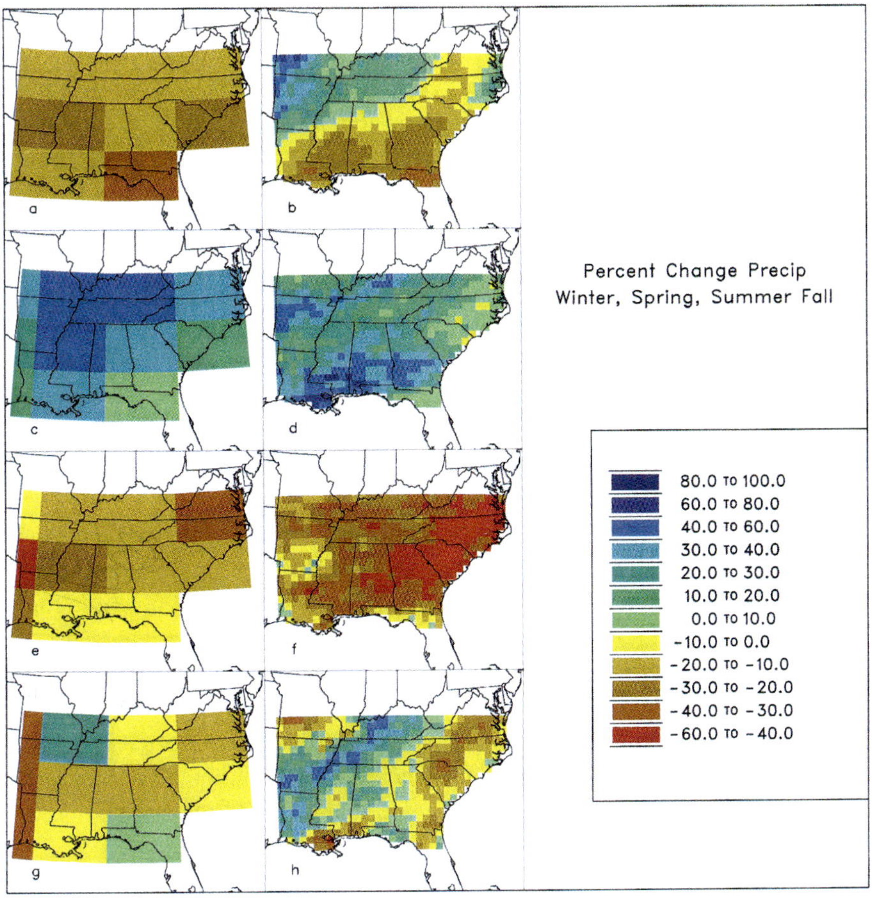

Figure 4. Percentage change ($2 \times CO_2$ – control/control) in seasonal precipitation for winter (a) CSIRO, (b) RegCM; spring (c) CSIRO, (d) RegCM; summer (e) CSIRO, (f) RegCM; and fall (g) CSIRO, (h) RegCM.

(Figures 4c,d). These increases are due to greater southerly water vapor fluxes over the region in the $2 \times CO_2$ simulation (not shown). However, differences between the spatial patterns of change in the CSIRO and RegCM are simulated. The CSIRO shows maximum increase of precipitation in the upper Mississippi Valley and Tennessee. The RegCM also shows a maximum of precipitation increase in the same region (although less extended than in the CSIRO), but in addition it simulates an area of large increase extending over the lower southcentral area of the region of interest (Figures 4c,d).

In summer, while both models show decreases throughout the area, the spatial patterns of these decreases are quite different (Figures 4e,f). Very large decreases in precipitation are found all along the eastern coastal plain for the RegCM (as much as 50%) whereas the changes in the CSIRO are more uniform across the domain with the largest changes over North Carolina and Virginia. Note that the region of maximum precipitation decrease in the RegCM corresponds to the region of maximum precipitation in the control run (see Figure 3h). Figure 6 shows the vertically integrated average summer water vapor fluxes (calculated from 6 hour samples) in the control and $2 \times CO_2$ experiments for both the CSIRO and the RegCM. Both models show in the $2 \times CO_2$ runs a minimum in water vapor flux occurring over the Gulf of Mexico off the coast of Florida. This causes a general decrease in southerly water vapor flow from the gulf regions to the southeastern U.S. which is most likely responsible for the simulated decrease in precipitation. Also note that in the RegCM this low in moisture flux is located farther north than in the CSIRO, which helps explain the greater precipitation decrease found in the regional model.

In winter the CSIRO experiences decreases in precipitation throughout the area (Figure 4a), while the RegCM experiences decreases in the southeastern half of the area, but precipitation increases across Arkansas and Tennessee (Figure 4b). Previous nested modeling experience has shown that, although the large scale features in the nested and driving models are generally similar, shifts in large scale circulations (e.g., storm tracks) can occur in the interior of the nested model domain (Jones et al., 1995; Giorgi et al., 1998). In our simulations the difference between CSIRO and RegCM precipitation change is due to a southerly shift of an area of increased precipitation that in the CSIRO only reached the central regions of Missouri and Kansas. Finally, in the fall (Figures 4g,h) both models show a decrease in precipitation over the southeastern coastal states and an increase in precipitation over the Upper Mississippi River Basin region included in the area of interest, although this increase extends further south in the RegCM.

For temperature the largest spatial contrasts in change are found in summer maximum temperature, when the RegCM exhibits large increases in temperature in the northeastern part of the domain (roughly corresponding to areas of the largest precipitation decrease), whereas the CSIRO exhibits uniform changes of between 4 and 5 °C throughout the area (Figures 5g,h). Changes in the RegCM vary from 3 to 4 °C in the southwest part of the area up to around 7 °C in the northeastern part of the study area. Interestingly, minimum temperatures in summer for the RegCM show rather uniform changes of around 4 °C across the study area (Figure 5f). A similar uniformity in change is seen for the CSIRO (Figure 5e). The large increases in maximum temperature for the RegCM in summer results from the extreme soil moisture drying that occurs because of the large decreases in precipitation. This results in large decreases in latent heat flux, and increases in sensible heat flux at the surface, and thus warming of the air above. Another notable contrast in temperature is seen in spring minimum temperature for the CSIRO (Figure 5a) when increases between 6 and 7 °C in the northern portion of the study area are seen. This increase is less pronounced in the RegCM (Figure 5b) and is largely lacking in the maximum temperatures (Figures 5c,d).

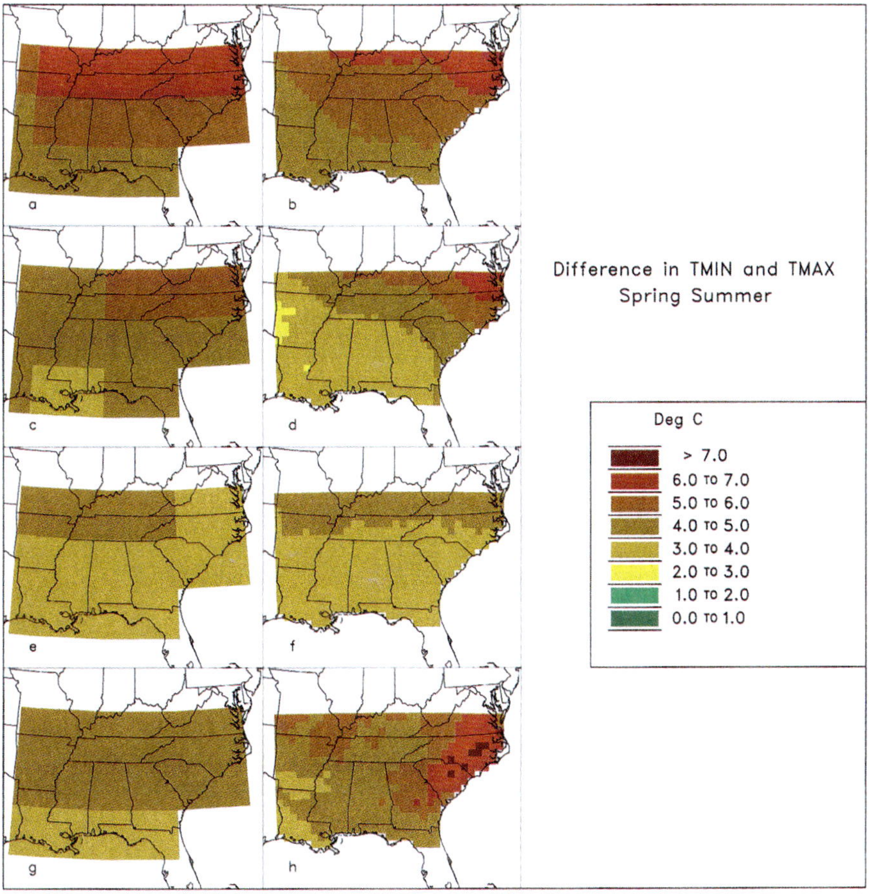

Figure 5. Changes ($2 \times CO_2$ – control) in minimum and maximum temperature (°C) for the CSIRO and RegCM for spring (a) CSIRO minimum, (b) RegCM minimum; (c) CSIRO maximum, (d) RegCM maximum; and summer (e) CSIRO minimum, (f) RegCM minimum; (g) CSIRO maximum, (h) RegCM maximum.

Temperature changes in the winter and fall are similar for both models (remaining around 4 or 5 °C) except in the fall maximum for the RegCM on the east coast, which resembles the pattern of summer (not shown).

4.2. STATISTICAL ANALYSIS OF MONTHLY CHANGES

The same statistical package used in comparing the control and observed precipitation and temperatures for both models, is used here to compare the $2 \times CO_2$ and control results. Table IIb presents the percentage of the domain found to have significant differences $2 \times CO_2$ – control in mean precipitation for each month for the two models.

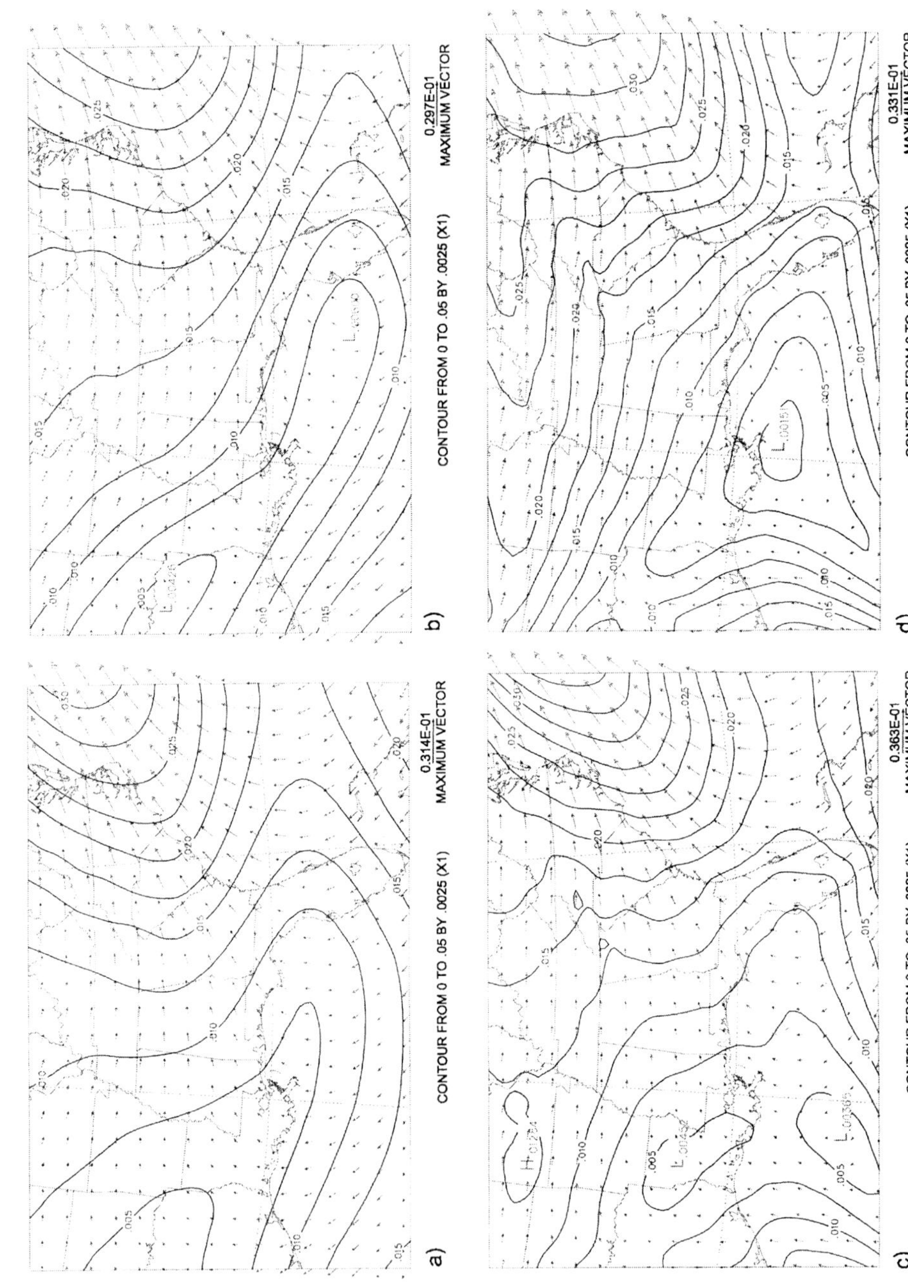

Figure 6. Vertically integrated moisture flux in summer: (a) CSIRO control, (b) CSIRO 2 × CO_2, (c) RegCM control, and (d) RegCM 2 × CO_2.

Again each CSIRO grid represents 12.5% of the southeast area considered. Based on the percentage area that exhibit significant changes, most significant decreases in the CSIRO simulation are seen in July through September, while most significant increases are seen in March and April.

For the RegCM, a similar pattern of significant decreases is seen in July through September, while the pattern of significant increases in the spring months differs from that of the CSIRO. In March and April the CSIRO shows much greater areas with significant increases, while the opposite occurs in May. In general the pattern of significant increases reflects the greater spatial variability and greater intensity of changes seen in the RegCM compared to the CSIRO.

All maximum and minimum temperature changes are significant for both models for all months (not shown). However, there are some interesting contrasts of degree of change on a monthly time scale. In particular in May in the RegCM rather small increases in temperature tend to dominate in the area where precipitation increases substantially in the southcentral subregion. This strong association between changes in temperature and precipitation in the RegCM is not as evident in the CSIRO.

4.3. FORMATION OF THE TWO CLIMATE SCENARIOS

To form the two scenarios to be used as input to crop models, the standard procedure of combining the change in climate from the climate models with the baseline observed climate data set was used. For maximum and minimum temperature, monthly average values of differences ($2 \times CO_2$ – control) were calculated. For the other variables (precipitation, solar radiation, relative humidity, and wind speed), the ratio $2 \times CO_2$/control for each month was calculated. These differences or ratios were then combined with the daily values of the observed data set. For example, the monthly difference in maximum temperature would be added to each daily value of observed maximum temperature. For temperature this procedure changes the mean of the temperature time series, but the variance (both daily and interannual) is unchanged. For precipitation, the daily values of observed precipitation are multiplied by the ratio for the particular month in question. Thus, the frequency of precipitation is not changed, but the mean of the daily intensity is changed by the ratio, and the variance of the intensity is changed by the square of the ratio. The two different resolutions of climate scenarios are produced based on the spatial resolution of the changes in climate. The changes in climate are combined with the baseline climatology with a resolution of 0.5 degrees. For the coarse (CSIRO) resolution scenario, then, the observed climatologies at all 50 km grids contained within a given CSIRO grid box are combined with the same set of climate changes (on a monthly basis). For the fine resolution (RegCM) scale scenario, each 50 km grid observed climate is combined with the unique set of climate changes from the RegCM grid corresponding to that location.

4.3.1. *Changes in Other Variables*

While this paper mainly concerns precipitation and temperature, we make a few comments here on the changes in some other variables (solar radiation, relative humidity, and wind speed) in the two different climate projections. All crop models applied in the various projects use solar radiation as an input, and some use relative humidity and wind speed. Changes in solar radiation are highly correlated with changes in precipitation, i.e., solar radiation increases where precipitation decreases. The changes in solar radiation, however, are relatively small, mostly within + or –10%.

In certain months, such as the summer months when both models experience large decreases in precipitation, increases in solar radiation can be up to 30%. The detailed spatial variability of these changes tend to be larger in the RegCM, but mean patterns across the domain are strikingly similar for the two models, in most months. Relative humidity changes also follow changes in precipitation to some degree. There are generally increases from February through June (on the order of 10%) and then decreases in July through October. In winter the CSIRO simulates slight decreases whereas the RegCM produces slight increases. Largest (about 20%) decreases occur in July and August in both models. Surface winds in general decrease (on the order of 10%) in both models in most months, but slightly larger decreases are found in the RegCM.

5. Signal-to-Noise Considerations

In this section we discuss two issues which are important for determining the robustness of the two different scenarios and whether they are truly different. One issue is the fact that only five years were used from the 30 year runs of the CSIRO to serve as boundary conditions for the RegCM. This is a relatively short time series, and certainly a more robust scenario could have been developed from using all 30 years of the CSIRO. This issue considers how representative the particular five years we used were in relation to the full 30 years of the CSIRO. A second issue is the signal-to-noise problem in determining how different the coarse scale and fine scale scenarios are from each other. Here the issue of sample size is also important. When these runs were initially produced for the long term agriculture project, limitations in computer time allocation prohibited us from producing longer runs. We demonstrate below how this affects the two issues.

5.1. THE FIVE YEARS OF CSIRO IN RELATION TO THE FULL RUN LENGTH

We compared the particular five years of control and $2 \times CO_2$ runs of the CSIRO we used with longer time periods to situate the particular five years of the climate scenarios (i.e., ratios or differences) in the context of the longer time series. In general we found that the longer the simulation, the smoother the seasonal cycle of the ratios (2 × CO_2/control) for precipitation (Figure 7). The five year simulations show greater variability of the seasonal cycle of the ratios (Figure 7), but the over all shape of the seasonal cycles are similar regardless of run length. This tendency was found at all grids. In this instance, in certain months, such as March, the ratios decrease with run length, but this is not uniformly the case. There are similarities in the ratios regardless of run length through the summer months, when decreases in precipiitation are found. The shortest run length does not have the most extreme ratio during these months, which are particularly important from the point of the crop model responses. Interestingly enough, at most grids the ratios in the spring months are higher for the five year segment than for the full 30 years, but ratios are very similar regardless of run length in the summer. From the point of view of agriculture, our 5-year CSIRO scenario likely is less draconian than the 30-year scenario, since the higher precipitation ratios (i.e., precipitation increases) in our scenario would result in less moisture stress in the spring, before the droughty summer months (when preciptation ratios are low), which occurs with all run lengths.

Using longer simulations of the CSIRO and then producing longer simulations of the RegCM thus would likely have produced different results than what we have produced here. Also, using a different five year period likely would have resulted in scenarios somewhat different from what we present here. However, since we are mainly interested in a sensitivity of the spatial scale effect of climate change scenarios, we are not overly constrained by this time slice problem. While a longer simulation would produce more robust results, and probably less extreme ratios in some months, our results are still useful as a sensitivity analysis. The difference between the five year ratios and 30 year ratios rarely resulted in different directions of change, but were generally a matter of degree.

Contrasts in the effect of run lengths for time periods within the full 30-year time series for temperature were much smaller than those for precipitation (Figure 8). At the same grid as in Figure 7, we see that the time series of control and doubled CO_2 for maximum summer temperature are highly stationary, resulting in little variation in the difference in the two time series with run length. The mean of the control run for the five years used in our scenario and the long-term mean was 31.5 °C, and the same for the 2 × CO_2 was 36.1 and 36.4 °C, respectively.

5.2. STATISTICAL COMPARISON OF THE FINAL SCENARIOS

Here we demonstrate that there are substantial differences in the final climate scenarios that serve as inputs to the crop models. We establish that the contrasts in the climate changes of CSIRO versus those of RegCM are statistically significant, and do not result from random noise. The only sensible way to do this, given the differences in the spatial scales of the scenarios, is to analyze the differences in the final scenarios, described in Section 4.3 above, that are produced on the same 50 km grid. These are the climate time series resulting from combining ratios and

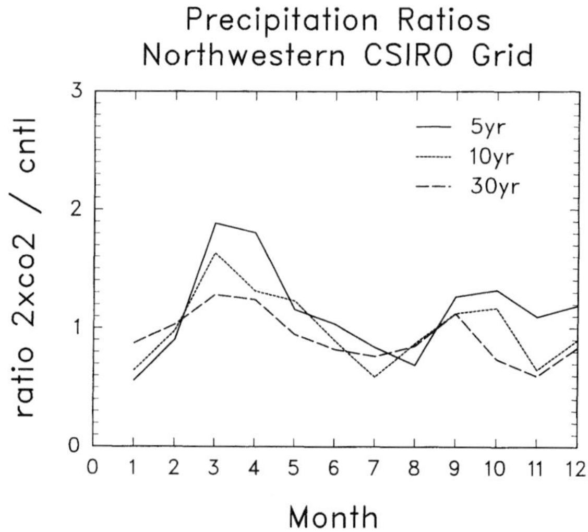

Figure 7. Ratios of precipitation 2 × CO_2/control for the CSIRO grid box in the northwestern part of the Southeast study area for time series of various lengths.

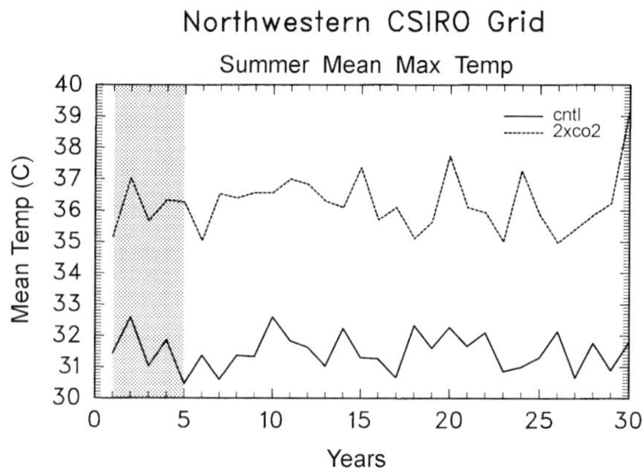

Figure 8. Thirty-year time series of summer mean maximum temperature from the CSIRO control and 2 × CO_2 runs, for the same grid box as in Figure 7 (northwestern part of the domain). Gray shading indicates the five years of the simulations used for the scenarios described in this paper.

differences with the observations. Each scenario contains 36 years of daily data at each 50 km grid point. We compared the final CSIRO scenario with the RegCM scenario applying the same statistical package used in Sections 3.2 and 4.2 above. We compared both the mean precipitation and mean temperatures. The most significant contrasts in the changed climate data sets for precipitation are in July (63% of grids are different at 0.05 level), August (60% significantly different, about 50% positive), September (33% significantly positive difference), and November (45% significantly negative, western part). Other months that have moderate amounts of significant differences are December (17% significant and negative); May (15%

significant and negative), March (13% significant mostly positive) and June (17% significant, mostly positive). In other months the areas of significant differences are under 10%. Temperatures are also significantly different in most months.

These results indicate that the differences in the changed climates produced as scenarios are substantial in many of the months important for agriculture in the Southeast. The fact that precipitation is substantially different in the summer months is particularly important. These results indicate that the contrasts in climate change from the two models are detectable as a signal. Even so, it must be emphasized that it is the total effect that the climate changes have on the crop models that ultimately matters. The crop models experience the complete integration of all the changes in the relevant variables for the entire year. Hence, establishing that the individual elements of the scenarios are statistically significantly different on a monthly basis is not the most crucial aspect of this study, from the point of view of impacts.

6. Summary and Conclusions

We have presented an analysis of equilibrium climate change experiments performed with a nested regional climate model over the southeastern United States and of the GCM experiments that provided the boundary conditions for the nesting. Our main conclusions can be summarized as follows:

1. Overall, in the control run the RegCM improved many aspects of the simulation compared to the driving CSIRO. In particular surface temperature and the spatial patterns of precipitation were better simulated in the regional model. The RegCM had a larger positive precipitation bias than the CSIRO in summer, which however was dominated by the contribution of a relatively small region.
2. The models reproduced the basic features of the climate of the region, but also showed significant deficiencies, most notably the failure to reproduce the spatial patterns of precipitation in the spring and fall, and a precipitation overestimation in summer. Minimum temperature was reproduced very well by the RegCM and overestimated by the CSIRO, while maximum temperature showed negative biases in both models, but these were smaller for the RegCM.
3. With doubled CO_2 forcing, both models simulated increases in temperature mainly between 4 and 5 °C, but larger increases were simulated for the spring minimum temperature by the CSIRO and for the summer maximum temperature by the RegCM. These changes are for the most part larger than the biases in both models, especially in the RegCM.
4. Both models simulated increases in precipitation in the spring, but these were larger for the CSIRO; and both models also simulated substantial decreases in summer precipitation, more so for the RegCM. These seasonal changes in precipitation were larger than the model biases in spring, but somewhat smaller than the biases in summer. The latter condition is common to other regional

Kowalczyk, E. A., Garratt, J. R., and Krummel, P. B.: *Implementation of a Soil-Canopy Scheme into the CSIRO GCM – Regional Aspects of the Model Response*, CSIRO Division of Atmospheric Research Technical Paper No. 32, CSIRO, Australia, 59 pp.

Legates, D. R. and Willmott, C. J.: 1990a, 'Mean Seasonal and Spatial Variability in Gauge-Corrected Global Precipitation', *Int. J. Clim.* **10**, 111–127.

Legates, D. R. and Willmott, C. J.: 1990b, 'Mean Seasonal and Spatial Variability in Global Surface Air Temperature', *Theor. Appl. Climatol.* **41**, 11–21.

Loveland, T. R., Merchant, J. W., Ohlen, D. O., and Brown, J. F.: 1991, 'Development of a Land Cover Characteristics Database for the Conterminous United States', *Photogrammetric Engineer. Remote Sens.* **57**, 1453–1463.

Mearns, L. O.: 2003, 'Issues in the Impacts of Climate Variability and Change on Agriculture. Applications to the southeastern United States', Guest Editorial, Special Issue *Clim. Change* **60**, 1–6.

Mearns, L. O., Giorgi, F., McDaniel, L., and Shields, C.: 1995a, 'Analysis of the Diurnal Range and Variability of Daily Temperature in a Nested Modeling Experiment: Comparison with Observations and 2CO Results', *Clim. Dyn.* **11**, 193–209.

Mearns, L. O., Giorgi, F., McDaniel, L., and Shields, C.: 1995b, 'Analysis of the Variability of Daily Precipitation in a Nested Modeling Experiment: Comparison with Observations and 2CO Results', *Global Plan. Change* **10**, 55–78.

Mearns, L. O., Hulme, M., Carter, T. R., Leemans, R., Lal, M., and Whetton, P.: 2001, 'Climate Scenario Development', Chapter 13, in Houghton, J. T., Ding, Y., Griggs, D. J., Noguer, M., van der Linden, P. J., Dai, X., Maskell, K., and Johnson, C. A. (eds.), *Climate Change 2001: The Scientific Basis, Contribution of Working Group I to the Third Assessment Report of the IPCC*, Cambridge University Press, Cambridge, pp. 739–768.

National Renewable Energy Laboratory (NREL): 1992, *National Solar Radiation Data Base (1961–1990)*, Users Manual, Version 1.0, Ashville, N.C., NOAA, NCDC, 93 pp.

New, M., Hulme, M., and Jones, P. D.: 1999, 'Representing Twentieth Century Space-Time Climate Variability. Part 1: Development of 1961–90 Mean Monthly Terrestrial Climatology', *J. Climate* **12**, 829–856.

New, M., Hulme, M., and Jones, P. D.: 2000, 'Representing Twentieth Century Space-Time Climate Variability. Part 2: Development of 1901–96 Monthly Grids of Terrestrial Surface Climate', *J. Climate* **13**, 2217–2238.

Pal, J. S., Small E. E., and Eltahir, E. A. B.: 2000, 'Simulation of Regional-Scale Water and Energy Budgets: Representation of Subgrid Cloud and Precipitation Processes within RegCM', *J. Geophys. Res.* **105**, 29579–29594.

Pan, Z., Christensen, J. H., Arritt, R. W., Gutowski, W. J. Jr., Takle, E. S., Otieno, F.: 2001, 'Evaluation of Uncertainties in Regional Climate Change Simulations', *J. Geophys. Res.* **106**, 17735–17753.

Richardson, C. W. and Nicks, A. D.: 1990, 'Weather Generator Description', in *EPIC – Erosion/Productivity Impact Calculator, 1*, Model Documentation, Agricultural Research Service, United States Department of Agriculture, Washington, D.C., 93–104.

Robinson, P. J. and Henderson, K. G.: 1992, 'Precipitation Events in the Southeast United States of America', *Int. J. Clim.* **12**, 701–720.

Takle, E. S., Gutowski, W. J., Arritt, R. W., Pan, Z., Anderson, C. J., Ramos da Silva, R., Caya, D., Chen, S.-C., Giorgi, F., Christensen, J. H., Hong, S.-Y., Juang, H.-M. H., Katzfey, J., Lapenta, W. M., Laprise, R., Liston, G. E., Lopez, P., McGregor, J., Pielke, R. A. Sr., and Roads, J. O.: 1999: 'Project to Intercompare Regional Climate Simulations (PIRCS): Description and Initial Results', *J. Geophys. Res.* **104**, 19443–19461.

Watterson, I. G.: 1998, 'An Analysis of the Global Water Cycle of Present and Doubled CO_2 Climates Simulated by the CSIRO General Circulation Model', *J. Geophys. Res.* **103**, 2311–23129.

Watterson, I. G., Dix, M. R., and Colman, R. A.: 1999, 'A Comparison of Present and Doubled 2 Climates and Feedbacks Simulated by Three General Circulation Models', *J. Geophys. Res.* **104**, 1943–1956.

Watterson, I. G., O'Farrell, S. P., and Dix, M. R.: 1997, 'Energy and Water Transport in Climates Simulated by a General Circulation Model that Includes Dynamic Sea Ice', *J. Geophys. Res.* **102**, 11027–11037.

(Received 2 May 2002; in revised form 17 March 2003)

THE EFFECT OF SPATIAL SCALE OF CLIMATIC CHANGE SCENARIOS ON SIMULATED MAIZE, WINTER WHEAT, AND RICE PRODUCTION IN THE SOUTHEASTERN UNITED STATES

E. A. TSVETSINSKAYA [1,2], L. O. MEARNS [2★], T. MAVROMATIS [3], W. GAO [4], L. McDANIEL [2] and M. W. DOWNTON [2]

[1] *Center for Remote Sensing and Department of Geography, Boston University, Boston, Massachusetts, U.S.A.*

[2] *The National Center for Atmospheric Research, 3450 Mitchell Lane, Boulder, CO 80301, U.S.A. E-mail: lindam@ucar.edu*

[3] *Department of Agricultural and Biological Engineering, University of Florida, Gainesville, Florida, U.S.A.*

[4] *Natural Resource Ecology Laboratory, Colorado State University, Fort Collins, Colorado, U.S.A.*

Abstract. We use the CERES family of crop models to assess the effect of different spatial scales of climate change scenarios on the simulated yield changes of maize (*Zea mays L.*), winter wheat (*Triticum aestivum L.*), and rice (*Oryza sativa L.*) in the Southeastern United States. The climate change scenarios were produced with the control and doubled CO_2 runs of a high resolution regional climate model and a coarse resolution general circulation model, which provided the initial and lateral boundary conditions for the regional model. Three different cases were considered for each scenario: climate change alone, climate change plus elevated CO_2, and the latter with adaptations. On the state level, for most cases, significant differences in the climate changed yields for corn were found, the coarse scale scenario usually producing larger modeled yield decreases or smaller increases. For wheat, however, which suffered large decreases in yields for all cases, very little contrast in yield based on scale of scenario was found. Scenario scale resulted in significantly different rice yields, but mainly because of low variability in yields. For maize the primary climate variable that explained the contrast in the yields calculated from the two scenarios is the precipitation during grain fill leading to different water stress levels. Temperature during vernalization explains some contrasts in winter wheat yields. With adaptation, the contrasts in the yields of all crops produced by the scenarios were reduced but not entirely removed. Our results indicate that spatial resolution of climate change scenarios can be an important uncertainty in climate change impact assessments, depending on the crop and management conditions.

1. Introduction

Many studies have considered the possible impacts of climate change on agriculture, in both global (Rosenzweig and Parry, 1994; Parry et al., 1999) and national or regional contexts (Adams et al., 1990; Alexandrov and Hoogenboom, 2000; Reilly et al., 2003). However, they are all confronted with numerous uncertainties, many of them associated with the uncertainties in predicting future climate. Two of the

★ Corresponding author.

Climatic Change **60:** 37–71, 2003.

main uncertainties concern the future paths of emissions of greenhouse gases and the response of different global climate models to a given external forcing (such as increased concentration of CO_2). For the most part agricultural impacts have explored the uncertainty related to the different regional projections of climate change produced by different climate models (e.g., Rosenzweig and Parry, 1994; Alexandrov and Hoogenboom, 2001; Tubiello et al., 2002; Reilly et al., 2003). However, an additional uncertainty that has plagued the impacts community for decades is associated with the spatial scale of the climate models and thus the information they produce. Impacts researchers have been dissatisfied by the relatively coarse resolution of information available from climate models (on the order of 100's of kilometers) (e.g., Gates, 1985). From the impacts perspective there is a serious scale mismatch between the coarse resolution climate model information and the typical scale of concern for regional impacts (one or two orders of magnitude finer scale). As the emphasis in climate change impacts analysis has continued to shift to the regional scale, the desire for more detailed high resolution climate change information has intensified.

There are now techniques available (e.g., statistical downscaling and regional climate modeling) whereby higher resolution climate change information can be developed (Giorgi et al., 2001); and these higher resolution results have been applied to impacts studies (e.g., Brown et al., 2000; Guereña et al., 2001; Thomson et al., 2002). Yet in very few of these applications has the sensitivity of the impacts results to the spatial scale of the scenarios been thoroughly examined. In other words, the value of the high resolution information has not been clearly determined. Mearns et al. (1999, 2001a) examined the effect of scenario spatial scale in the Great Plains of the U.S. for corn, wheat, and soybeans and found the scale of the scenarios significantly affected the results in terms of percentage change of yields.

In this study we investigate further the uncertainty introduced into climate change agricultural assessments due to varying spatial scale of climate change information. We consider corn, wheat, and rice production in the southeastern United States for a variety of management conditions and exercise them with high and low resolution climate scenarios.

We use the Crop Estimation through Resource and Environment Synthesis (CERES) family of crop models (i.e., CERES-Maize (Hoogenboom et al., 1994; Jones and Kiniry, 1986), CERES-Wheat (Ritchie and Otter, 1985), and CERES-Rice (Godwin et al., 1992)). We evaluate all three crop models for their ability to reproduce mean and variability of yields at a variety of experimental station locations in the Southeast. This we consider important particularly because changes in variability of crop yields can also be an important aspect of future agricultural production that also can affect agricultural economics.

We then simulate crop responses to climate change scenarios produced with a coarse-resolution general circulation model (GCM) and a fine resolution regional climate model (RCM). The CERES family of crop models has been widely used to assess climate change impacts on the agricultural systems of various regions throughout the world (Aggarwal and Mall, 2002; Curry et al., 1990; Dhakhwa et al., 1997; Mearns et al., 1996, 1999; Southworth et al., 2000, 2002; Reilly et al., 2003).

We consider three cases for each climate scenario: the effects of climatic change only, combined climatic change and elevated CO_2 effects, and the above plus effects of adaptations to climatic change. Changes in yields from base conditions for the two climate scenarios are investigated at multiple levels of aggregation. In particular we consider the state level of aggregation, since this is the scale of changes in crop yield that is used in another part of the over all project, in the study on the economic effects of the crop yield changes (Adams et al., 2003, this issue). Since the economic model used requires crop yields throughout the U.S., we also calculated in less detail than for the Southeast changes in crop yields for the rest of the U.S. outside the Southeast, and briefly report those results. We also investigate which aspects of the contrasts in the climate scenarios at the two scales account for the differences in the yields simulated using the scenarios as input.

2. Materials and Methods

2.1. CERES CROP MODELS

Crop models used in our study are version 3.1 of CERES-Maize, CERES-Wheat, and CERES-Rice. Processes simulated by these models include crop phenological development, biomass production and allocation among plant parts (including roots, stems, leaves, and grain), and senescence of leaves. The models simulate both the potential biomass production and the water-, temperature-, and nutrient-limited production. Direct effects of CO_2 on photosynthesis and water use are also simulated. CERES-Rice incorporates additional processes, such as the effect of transplanting on the growth and development of rice, and nitrogen and water dynamics under flooded, intermittently flooded, and upland conditions. The CERES models require daily weather inputs, soil parameters at various profile depths, and management details. Also required are cultivar-specific genetic coefficients, which determine the duration of a given growth stage, response to photoperiod, and seed characteristics. For specifics of model parameterizations we refer the reader to Hoogenboom et al. (1994), Jones and Kiniry (1986), Ritchie and Otter (1985), and Ritchie et al. (1986).

2.2. CROP MODEL EVALUATION

All CERES models have undergone extensive evaluations. Earlier studies in the United States found that the CERES-Maize model accurately reproduced maize grain yield in the U.S. corn belt (Hodges et al., 1987), county yield estimates in nine U.S. states (Kiniry et al., 1997), and yield and nitrogen uptake in California (Pang et al., 1997). CERES-Maize has also been extensively tested outside the U.S., e.g., in Brazil (Liu et al., 1989), China (Wu et al., 1989), and Zimbabwe (Phillips et al., 1998). The CERES-Wheat model also has been tested in diverse environments, e.g., in Canada (Moulin and Beckie, 1993), Hungary (Bacsi and Hunkar, 1994), the U.S.A. (Rosenzweig, 1990; Kiniry and Bockholt, 1998; Tubiello et al., 1999). The CERES-Rice model has been widely evaluated as well (Cheyglinted et al., 2001; Mall and Aggarwal, 2002).

Most of the earlier studies with the CERES model focused on mean yield rather than its interannual variability. The year-to-year variability in yield associated with the interannual climate variability has been explored to a much lesser degree (Mearns et al., 1992; Kiniry and Bockholt, 1998; Phillips et al., 1998). We specifically evaluate, using yield performance trials data, the CERES (-Maize, -Wheat, and -Rice) models for their ability to reproduce both the mean and the year-to-year variability in measured yields over series of years throughout the Southeastern United States. To test the CERES models' performance, crop, soil, and management data were collected from crop variety performance trial bulletins available from agricultural experiment stations in the Southeast (Table I). Our strategy was similar to that of Kiniry and Bockholt (1998), who evaluated the performance of the CERES and ALMANAC models against yield performance trials data in Texas. For a wide range of cultivars and management conditions, we tested several model aspects, including crop life cycle (anthesis and physiological maturity dates) and yield (yield components and final harvested yield).

For corn, anthesis and harvest dates were recorded for each treatment. We assumed the time from physiological maturity (growth stage simulated by CERES-Maize) to harvest (recorded from the field data) for corn to be four weeks for medium season varieties (three weeks for short season varieties) in Georgia (pers. comm., P. L. Raymer, 1998) and 15 days in Louisiana (Mascagni et al., 1998). For most maize treatments, grain yield at harvest (kg/ha), individual ear grain weight (g), and number of ears per plant were measured. For most winter wheat treatments, heading (or ear emergence) was the only growth stage recorded. Since CERES-Wheat simulates anthesis, and not heading (which is recorded in the field), we assumed anthesis to occur eight days after heading (McMaster et al., 1992). Grain yield at harvest and harvest date were also recorded for winter wheat. Maturity was assumed to occur five to seven days before harvest, depending on the location. Our results could be to some degree affected by using the Julian rather than thermal time to infer maturity dates from harvest dates for maize and winter wheat. However, the agricultural experts we consulted believed that using a constant number of days would be adequate. For rice, planting, heading, and maturity dates were recorded, as well as final grain yield (at 12% moisture content).

Daily maximum and minimum air temperature, precipitation, and solar radiation from weather stations at or near the experimental stations were used to drive the CERES crop models. See Mearns et al. (2003b, this issue) for details on the climate data sets. Soil types were taken from the experimental station data (Table I), but requisite soils characteristics were determined from the State Soil Geographic

Table I

List of agricultural experimental stations in the Southeast, where the CERES crop models were evaluated. Also listed are specifics of setup pertaining to crop, soil, and management. Planting depth was set to 5 cm for corn and winter wheat and 2.5 cm for rice

Crop	Experimental sites	Years of analysis	Planting date	Planting density	Row spacing	Fertilizer amount	Soils
Corn	Tifton, GA (31.47° N, 83.52° W)	1986–96	mid-March to mid-April	5.6–6.7 plants/m ^{2}	0.76–0.1 m	146–292 kg/ha (2 applications)	Deep sandy loam
	Alexandria, LA (31.32° N, 92.47° W)	1988–95				140–270 kg/ha (2 applications)	Norwood silty loam
	Baton Rouge, LA (30.37° N, 91.16° W)						Commerce silty loam
	Jeanerette, LA (29.95° N, 92.72° W)						Baldwin silty clay
	St. Joseph, LA (31.95° N, 91.23° W)						Deep Sharkey clay, medium Commerce silty loam
	Winnsboro, LA (32.52° N, 92.05° W)						Medium depth Gigger silty loam
Winter wheat	Simpson, SC (34.68° N, 82.82° W)	1991–96	Early October mid-December	324–330 plants/m ^{2}	0.18–0.25 m	90–120 kg/ha (2 applications)	Shallow sandy loam
	Pee Dee, SC (34.18° N, 79.44° W)						Deep loamy sand
	Edisto, SC (33.37° N, 81.32° W)						Deep loamy sand
	Crossville, AL (34.28° N, 85.97° W)	1989–94					Deep fine sandy loam
	Prattville, AL (32.48° N, 86.48° W)						Sandy loam
	Fairhope, AL (30.55° N, 87.88° W)						Sandy loam
Rice	Crowley, LA (30.00° N, 90.50° W)	1983–92	Early April to mid-May	150 plant/m ^{2}	0.2 m	80–280 kg/ha (2–3 applications)	Crowley silt loam
	Stoneville, MS (33.25° N, 90.55° W)	1984–95					Sharkey clay
	Lambert, MS (34.20° N, 90.30° W)						Forestdale silty clay loam

(STATSGO) data base (USDA, 1994). Soil parameters (e.g., drained upper limit) were estimated from the soil characteristics using the Baumer et al. (1994) method. The Priestley–Taylor method (Ritchie, 1985) was used to compute potential evapotranspiration. For maize and winter wheat, initial soil water and nitrogen amounts at planting time were estimated by running the crop models in sequence mode starting January 1 four years prior to the first growing season, assuming a rainfed continuous rotation. For rice, soils were assumed to be saturated at initial planting, and the depth of the flood water (at planting) was obtained from experimental data. For irrigated maize treatments, since only the total irrigation amount and the number of applications were available from the field record, we allowed CERES-Maize to specify the timing of irrigation (i.e., an automatic irrigation option was used). No irrigation water was applied at any of the winter wheat sites. Rice was flushed periodically until the permanent flood was established (i.e., 4–5 weeks after planting, 10 cm water depth). Automatic irrigation was used to minimize the effect of water shortage on rice yield production. Weeds, diseases and insect pests, excessive water, wind damage, and limiting nutrients other than nitrogen were not considered in our simulations.

A randomly-selected half of the experimental data sets were used in estimating the genetic coefficients and the remaining half were used for model evaluation. This approach is consistent with previous studies (Iglesias, 1995; Wu et al., 1989; Hodges et al., 1987). The coefficients were first calibrated in relation to phenology based on the thermal time integrals. Once the phenology coefficients (and therefore the simulated number of days available for grain filling) were calibrated, the yield component coefficients (G2 and G3) were adjusted to represent as accurately as possible grain components and the harvested yield. We tested the goodness of fit between the simulated and measured anthesis dates, maturity dates, and grain yield at harvest. For wheat and corn, we developed a single set of genetic coefficients each that allowed for the best agreement between the modeled and measured crop yields and timing of phenological development (Tables II and III). The wheat cultivar Arthur was used as the starting point for estimating the coefficients for the medium season cultivar for Florida 302. An additional wheat cultivar, Gore, was evaluated in Alabama, but was not used in the larger southeast study. For rice, the early season semi-dwarf variety Lemont from the DSSAT data base was used with only very minor adjustments to coefficients. We later used these genetic coefficients in a domain-wide study of the climate change impacts on crop production in the southeastern United States (Section 2.4). This approach is supported by the results of Hodges et al. (1987) and Iglesias (1995), who found that a single set of coefficients can be representative of a group of cultivars with similar characteristics grown in a particular geographic area.

Table II

Genetic coefficients for maize used in the Southeastern United States domain

Cultivar	P1	P2	P5	G2	G3	PHINT
Georgia Medium	200	0.2	990	700	8.2	75

P1 is the thermal time from seedling emergence to the end of the juvenile phase; P2 is a development delay factor for hours when photoperiod exceeds the threshold for maximum development rate; P5 is the thermal time from silking to physiological maturity; G2 is the maximum possible number of kernels per plant; G3 is the kernel filling rate (mg/day); PHINT is the phylochron interval.

Table III

Genetic coefficients for winter wheat used in the Southeastern United States domain

Cultivar	P1V	P1D	P5	G1	G2	G3	PHINT
FLORIDA302	6	2.7	5.2	3.4	1.8	1.9	95

P1V, P1D are the relative amounts that development is slowed for each day of incomplete vernalization (V) or when photoperiod is less than optimum (D); P5 is the relative grain filling duration based on thermal time; G1 is the kernel number per unit weight of stem; G2 is the kernel filling rate under optimum conditions; G3 is the non-stressed dry weight of a single stem and spike when elongation ceases (g); PHINT is the phylochron interval.

2.3. CLIMATE SCENARIOS

A present-day baseline climate data set and two climate change scenario data sets were constructed for the study region, which is comprised of nine states (Alabama, Arkansas, Florida, Georgia, Louisiana, Mississippi, North Carolina, South Carolina, and Tennessee). The present-day (1960–1995) baseline climate data set was compiled from a network of about 500 cooperative weather stations and gridded on a 50 km equal area grid (the grid of the regional model). Each grid box was assigned a single weather station nearest to its center. The baseline data set included daily incident solar radiation, maximum and minimum air temperature, and precipitation, and served as the base for all scenarios. See Mearns et al. (2003b) for more details on the baseline and climate change scenarios.

A coarse resolution climate change scenario was produced from climate simulations with the Australian Commonwealth Scientific and Industrial Research Organization (CSIRO) global circulation model (hereafter referred to as the General Circulation Model, or the GCM). A fine resolution climate change scenario was produced from runs with the National Center for Atmospheric Research Regional Climate Model version 2 (hereafter referred to as the Regional Climate Model, or the RCM), which was nested in the CSIRO GCM. Both the GCM and

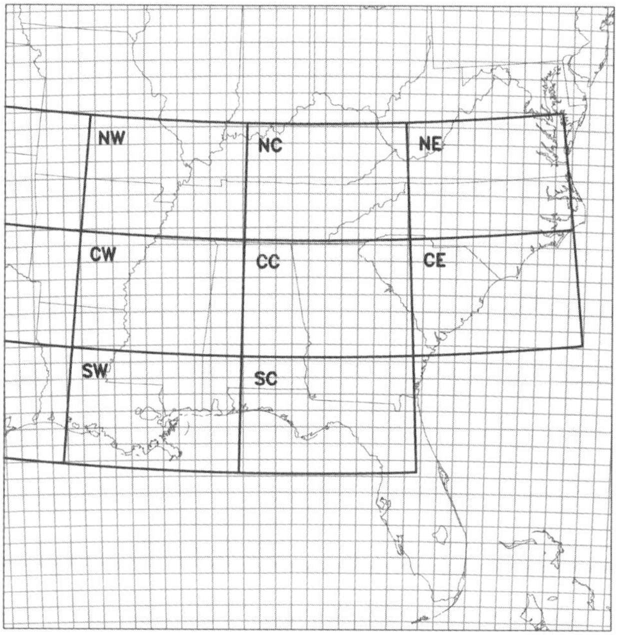

Figure 1. Map showing main study area. Heavy black lines indicate GCM grid boxes, lighter-lined small squares indicate RCM grid boxes (and state borders). The main 8 GCM grid boxes discussed in results are labeled with their relative locations. The three additional unlabeled GCM grids to the west were also used for climate scenario formation for the western edge of the study area. NW = northwest; NC = north central; NE = northeast; CW = central west; CC = central; CE = central east; SW = southwest; SC = south central. The scale of the map can be inferred from the fact that the RCM grids have a width of about 50 km.

the RCM were run under the current CO_2 conditions (330 ppmv CO_2) and the doubled CO_2 conditions (660 ppmv CO_2), assumed to be equivalent to the total effect of all greenhouse gases. The spatial resolution of the RCM was 50 km (414 RCM grid boxes covered the southeastern domain) and that of the GCM was 3.2° latitude by 5.6° longitude (11 GCM grid boxes covered the study domain). Figure 1 shows the GCM and RCM grid boxes over the study area. A detailed description of these climate data sets can be found in Mearns et al. (2003b). Climate change scenarios were constructed by applying monthly differences (for temperature) and ratios (for solar radiation and precipitation) between the $2 \times CO_2$ and control climate simulations to the present-day baseline climate data set. For RCM scenarios, a unique monthly value was applied to each of the RCM grids. For GCM scenarios, eleven unique deltas/ratios covered the southeastern domain (i.e., all RCM grids within one GCM grid box received the same deltas/ratios). In these scenarios we do not take into account changes in daily and interannual variability that might have occurred in the climate model runs. Variability is inherited from the baseline climatology.

An observed climate data set and a set of coarse and fine resolution climate change scenarios were also compiled for the rest of the conterminous United States. The observed data set consisted of stations selected to represent the various important agricultural regions in the rest of the U.S., but with a coarser spatial coverage than what was constructed for the Southeast. These scenarios were used as input to the CERES models to estimate yield changes for the major U.S. cropping areas. These yield changes were needed as input into an economic model (Agricultural Sector Model, ASM (McCarl et al., 2000)). See Mearns et al. (2003b) and Adams et al. (2003) for a discussion of the economic results. The same control and doubled CO_2 runs of the GCM and the same version of the RCM as for the southeastern scenarios were used to produce this set of scenarios. Results from two additional RCM runs were used to produce the fine resolution climate change scenario for the rest of the conterminous U.S., one for the western two-thirds of the U.S. (Giorgi et al., 1998), and one for the Great Lakes region (Bates et al., 1996, and Bates et al., unpublished data). Note that the Great Lakes are not represented in the GCM (due to the model's coarse spatial resolution) but are in the RCM.

2.4. SOUTHEAST REGIONAL ASSESSMENT APPROACH

We tested the effect of the GCM and RCM climate scenarios in the crop models for three different future climate cases: (a) climate change only (CO_2 concentration for the crop models at 330 ppmv); (b) climate change plus CO_2 fertilization effect (CO_2 concentration at 540 ppmv); and (c) case (b) plus adaptation to climate change. A base case was also produced where the crop models used observed climate data and CO_2 concentration of 330 ppmv. Note that for the CO_2 fertilization and adaptation cases, CO_2 amount is not the same as the equivalent doubling of CO_2 used in the climate model simulations (660 ppmv). In reality for an equivalent CO_2 doubling, concentrations of a number of different greenhouse gases increase. We calculated the portion of the doubling that would actually be CO_2, based on a common emissions scenario, IPCC IS92a, wherein equivalent CO_2 increases by about 1% per annum (IPCC, 1996). In that greenhouse gas emissions scenario, at the point of equivalent CO_2 doubling, the actual CO_2 concentration would be 540 ppm, not 660 ppm.

The resultant seven crop/climate scenarios/cases are: (1) *baseline case*, in which the baseline climate data set is used and the ambient CO_2 concentration of 330 ppmv is assumed (for the crop models); (2) *GCM climate change only case*, where GCM deltas/ratios are used to construct the climate change scenario and the ambient CO_2 concentration of 330 ppmv is assumed; (3) *RCM climate change only case*, where RCM deltas/ratios are used to construct the climate change scenario and the ambient CO_2 concentration of 330 ppmv is assumed; (4) *GCM elevated CO_2 case*, where GCM deltas/ratios are used and the ambient CO_2 concentration of 540 ppmv is assumed; (5) *RCM elevated CO_2 case*, where RCM deltas/ratios are used and the ambient CO_2 concentration of 540 ppmv is assumed; (6) *GCM adap-*

tation case, which is same as (4) plus adaptations to climate change; and (7) *RCM adaptation case*, which is same as (5) plus adaptations to climate change. For cases (6) and (7), the incorporated adaptations were changes in sowing dates, which were determined independently for GCM and RCM cases. Changes in cultivars were also attempted as an adaptation. However, changes in cultivars that we tested were not successful. These standard adaptation techniques have traditionally been used in climate change agricultural assessment studies.

Under each of the seven scenarios/cases, annual yields (in kg/ha) were computed for each RCM grid box over the entire Southeast for corn and wheat for 36 and 35 growing seasons, respectively. Rice yields were only simulated along the Mississippi river valley (a subdomain of 83 RCM grid boxes that spans from Arkansas to Louisiana and Mississippi). A single set of genetic coefficients obtained during the crop model evaluations was used for each of the three crops throughout the southeastern United States (Tables II, III). For corn and winter wheat we used state-specific planting windows (Table IV), based on the United States Department of Agriculture (USDA) National Agricultural Statistics Service (NASS) data (http://www.usda.gov/nass/pubs/uph97.htm). We therefore allowed planting dates to vary from year to year in response to season-specific thermal and moisture conditions. Note that while planting windows were the same across the five cases (all but the adaptation cases), the actual planting dates varied slightly among the cases in response to scenario-specific temperature and moisture conditions. For rice three single planting dates were used: day of year (DOY) 110 for the northern tier of grids, DOY 105 for the central tier of grids, and DOY 100 for the southern tier of grids, for all cases except for the adaptation runs. These dates were based on estimates from the NASS data and the experimental field trials. To determine sowing dates for the adaptation cases, we iteratively calculated crop yields for sowing dates that ranged from one month ahead of the base sowing date to one month after the base sowing date (e.g., Figure 2 for corn), and picked the date that insured the highest yields on a state level (regional level for rice). For all crops and all cases, harvesting was assumed to occur when the crop reaches maturity.

For corn, winter wheat, and rice, for the entire Southeast study region we assumed uniform plant population densities of 6.7 plants/m^2, 324 plant/m^2, and 150 plants/m^2, respectively. Row spacing was assumed to be 76 cm, 18 cm, and 20 cm, respectively. Planting depth was assumed to be 5 cm for corn and winter wheat, and 2.5 cm for rice. These parameter values fall within the ranges used for crop model evaluation (Section 2.2, Table I) and were year- and scenario-invariant.

Soil series from the STATSGO data base (USDA, 1994) were gridded on the RCM 50 km equal area grid. We determined soil physical properties of the best agricultural soil in the grid box and used these to represent the soil characteristics for the grid. All the crop model simulations described in this paper used the best agricultural soil series. The drained upper limit (DUL) and the lower limit (LL) and other soil properties were estimated for each soil series using the methodology

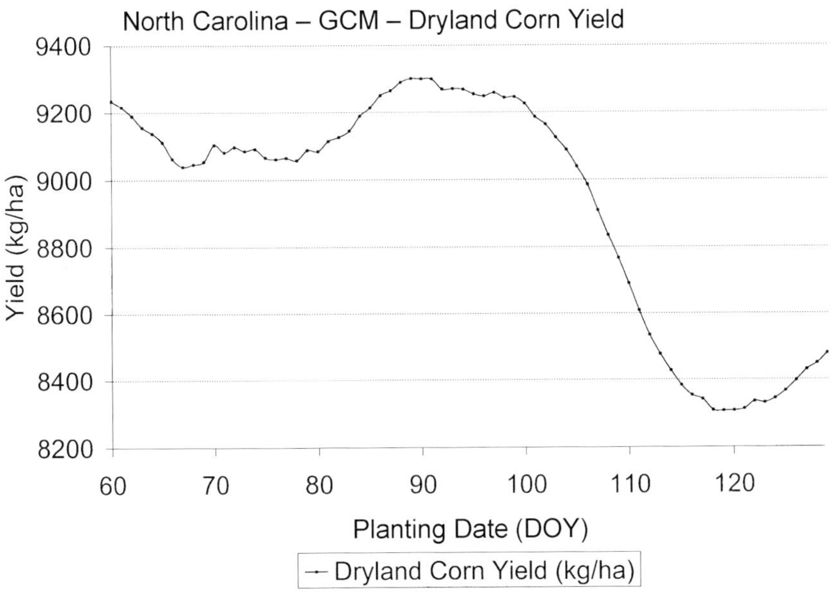

Figure 2. Yield response to varying sowing dates for corn for North Carolina under the GCM climate scenario. Original planting date (without adaptation) was around day 109 (April 17). The new planting date selected to optimize the yield was approximately day 89 (March 30).

Table IV

Most active planting dates for corn and winter wheat

State	Planting window	
	Corn	Winter wheat
Alabama	Mar 25–Apr 25	Oct 21–Nov 21
Arkansas	Apr 10–May 18	Oct 17–Nov 12
Florida	Mar 15–Apr 15	Nov 15–Dec 15
Georgia	Mar 20–Apr 15	Nov 15–Dec 5
Louisiana	Mar 19–Apr 4	Oct 30–Nov 21
Mississippi	Mar 31–Apr 28	Oct 8–Nov 12
North Carolina	Apr 10–Apr 25	Oct 15–Nov 20
South Carolina	Mar 20–Apr 20	Nov 11–Dec 15
Tennessee	Apr 15–May 1	Oct 15–Nov 10

Based on the United States Department of Agriculture (USDA) National Agricultural Statistics Service (NASS) data. These were used as the planting windows described in the text.

of Baumer et al. (1994), based on soils characteristics (e.g., percentage silt, clay, sand). The Priestley–Taylor method (Ritchie, 1985) was used to compute potential evapotranspiration. Dryland and irrigated runs were produced (irrigated only runs for rice) and no nitrogen stress was assumed.

2.5. CONTERMINOUS UNITED STATES (EXCLUDING THE SOUTHEAST)

Percentage changes in crop yields (irrigated and/or dryland) were produced for all 63 primary spatial units (mostly states) of the Agricultural Sector Model (ASM). Yield changes were based on the GCM and RCM climate change scenarios, for the elevated CO_2 and adaptation cases, and served as input into the ASM. Corn (continuous rotation, dryland and/or irrigated) was simulated in 31 states outside the Southeast, winter wheat (continuous and/or fallow rotation, dryland and/or irrigated) in 32 states, and spring wheat (continuous and/or fallow rotation, dryland and/or irrigated) in 5 states. Outside the Southeast, rice (continuous rotation, irrigated) was simulated in California, Missouri, and Texas.

Planting windows for each crop and each state were set based on the United States Department of Agriculture (USDA) National Agricultural Statistics Service (NASS) data (http://www.usda.gov/nass/pubs/uph97.htm). For corn for each state, an individual cultivar was set based on Kiniry et al. (1997) and our Southeast crop model evaluation results. For winter wheat, an individual cultivar was set for each state based on Mearns et al. (1997) and the USDA NASS state statistical offices data (http://www.usda.gov/nass/sso-rpts.htm). For spring wheat, Butte cultivar was used in all the relevant states. For rice, we used the same cultivar in Missouri and Texas as in the southeastern United States, and Cypress cultivar was used for California.

2.6. STATISTICAL TESTS ON THE SIGNIFICANCE OF THE SIMULATED YIELD CHANGES

For each crop, we performed two sets of comparisons. We tested (1) whether yields produced under each climate change scenario were significantly different from those produced under the base case, and (2) whether the yields produced in the RCM cases were significantly different from those produced in the GCM cases. We employed linear modeling techniques to test the hypothesis that mean yields from *all* the scenarios were equal. If the hypothesis was rejected, then pairwise comparisons were performed to test whether selected pairs of scenarios produced significantly different mean yields. Standard analysis of variance (ANOVA) methods require that elements of the sample data be independent; however, the simulated yields are spatially correlated. To take the spatial autocorrelation into account, we use a mixed models approach (Littell et al., 1996).

In fitting the mixed models, a model of the spatial correlation is estimated first, then mean yields are estimated for each scenario (a fixed effect) and each year (a random effect), along with their standard errors. Adjusted mean yields (also known as 'least squares means') are computed, representing mean yields which have been adjusted for spatial correlation and differences in mean yields between years. These adjusted means and their standard errors are used to determine whether there are statistically significant differences in mean yields between the seven climate cases (see Section 2.4 for a list of the seven climate cases). Means comparison models were fit in this way for the entire Southeast (all 414 RCM grids), for each of the nine states in the region, and using the unit of the GCM grid box.

For each crop we performed a regression analysis to determine what aspects of the contrasts in climate scenarios were most responsible for the differences in model yields under the two climate scenarios. A separate analysis was performed for each GCM grid box (i.e., the 50 km grids contained within each of the labeled GCM grid boxes in Figure 1). This analysis was performed on the GCM grid box level because it is that unit which determines the spatial variability of the GCM-based climate change cases. For each RCM grid box, we computed monthly means (over 36 seasons (35 seasons for winter wheat)) of each of the climate variables and the yields for each climate change scenario and then took the difference between the means (GCM minus RCM). The focus was on explaining the spatial variability of the differences in the future climate yields. Climate variables were only considered for the months between planting and harvest. We used an 'all possible regressions' approach, which finds the best possible model fit (based on maximum R^2) that can be obtained using a particular number of predictors. Models containing different number of predictors were compared using adjusted R^2 and Akaiki's Information Criterion (AIC). Note that additional crop model runs using uniform soils across the Southeastern domain were used for this analysis so that the spatial variability of the soils did not confound the analysis (i.e., we determined the most common soil in the study region and used it for all the 50-km grids for all cases to simulate yields and then used those yields in the multiple regression analysis). The percent changes in the yields from the baseline using the uniform soil were very similar to those produced with the cases using spatially varied soils.

3. Results and Discussion

3.1. MODEL SITE EVALUATION

The CERES-Maize model reproduced accurately the measured anthesis dates over an eleven-year period at six locations in Georgia and Louisiana. Both the mean and the year-to-year variability in anthesis dates were captured well (Table V). Physiological maturity was simulated well (16 location/years), but with lower accuracy than anthesis, as indicated by the root mean square errors (RMSE). This may in part be due to the fact that we used Julian rather than thermal time to infer maturity from recorded harvest dates.

The CERES-Maize model closely matched the mean of the measured yields for irrigated and rainfed treatments at Tifton, (within 4% of the measured yields),

Table V

Summary statistics for validation of CERES-Maize for medium season maize

$n = 8$	Anthesis (days)		Maturity (days)		Ear grain weight (g)		Grain yield (kg/ha)	
	Observed	CERES-Maize	Observed	CERES-Maize	Observed	CERES-Maize	Observed	CERES-Maize
a. Tifton, Georgia								
Mean	74.7	76.0	131.0	129.4	186	135	8767	8394
Std	9.2	10.8	16.6	11.5	31	37	1795	2606
Cor.		0.95		0.96		0.57		0.77
RMSE		3.5		6.2		59		1622

$n = 9$	Anthesis (days)		Maturity (days)		Grain yield (kg/ha)	
	Observed	CERES-Maize	Observed	CERES-Maize	Observed	CERES-Maize
b. Sites in Louisiana						
Mean	74.2	74.8	134.6	132.0	6417	6593
Std	6.4	8.3	9.6	7.0	1648	1380
Cor		0.83		0.80		0.80
RMSE		4.3		6.1		956

Std = Standard Deviation.
Cor = Correlation.
RMSE = Root Mean Square Error.

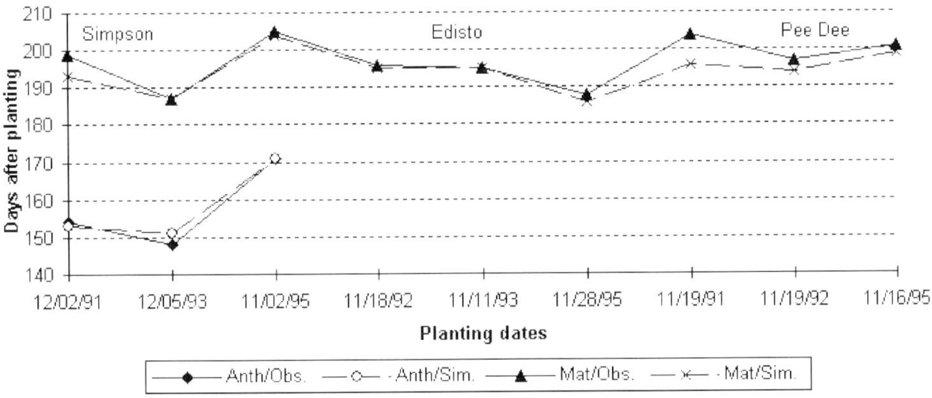

	Anthesis date		*Maturity date*	
	Observed	**Simulated**	**Observed**	**Simulated**
Mean	158	158	197	194
Std	12	11	6	6
Cor	0.99		0.90	

Figure 3. Evaluation of CERES-Wheat simulated anthesis and maturity dates in South Carolina.

but overestimated the year-to-year variability of yield. It closely matched mean yield (within 2% of observations), but underestimated variability of yield at the Louisiana locations (Table V(b)). Our results confirm the conclusion of Kiniry et al. (1997) that a model, which simulates grain yield reductions based largely on drought stress (i.e., CERES-Maize) is limited in its ability to explain the year-to-year variability at a location with consistently adequate rain (i.e., the Louisiana sites).

CERES-Wheat *cv.* Forida302 estimated very well anthesis dates, in terms of both the multi-year mean and the year-to-year variability (Figure 3). Maturity was simulated somewhat early, with the interannual variability somewhat lower than that observed. In only two out of nine treatments, however, the difference between the modeled and observed maturity dates exceeded six days. A close relationship ($r = 0.78$) between the measured and simulated grain yield was found, and the model overestimated mean observed yield by only 8% (Figure 4). The year-to-year variability of the simulated yield, however, was somewhat lower (14%) than the observed. CERES-Wheat overestimated the actual yields in 8 out of 9 treatments (Figure 4).

While the agreement between the multi-year mean CERES-Wheat simulated and observed yields (and anthesis and maturity) was generally good, during several seasons large discrepancies occurred. For example, during the 1995–96 growing season in Simpson, CERES-Wheat reproduced both anthesis and maturity dates very well (Figure 3), but overestimated yield by 36% (Figure 4). While rainfall

spatial variability of the changes is naturally much greater in the RCM, which also experiences greater extremes of changes.

Temperature increases in the GCM generally remain in the range of 3 °C to 5 °C, the higher values dominating in winter and spring. However, high (6 °C to 7 °C) increases in minimum temperature are found in March and April. In the RCM, temperature increases are much more spatially and temporally variable, ranging from 1 °C to 7 °C across seasons. In summer, the RCM simulates large increases (up to 7 °C) in maximum temperature in the northeastern part of the domain where maximum drying occurs, whereas the increases in the GCM are smaller. In spring (May) small increases (1 °C) dominate in the southcentral area where precipitation substantially increases.

We also provide a brief description of climate change scenarios for the major U.S. cropping areas outside the Southeast study region. In the Great Plains and the western Midwest, both the GCM and the RCM runs simulate precipitation increases during most seasons, which are particularly large in spring and continue through July. During these important cropping months, the increases in precipitation are generally greater for the RCM. Temperature increases, which are generally between 4 °C and 5 °C, are somewhat higher in the RCM than the GCM in spring, and are similar between the two models during summer. In the Southern Great Lakes region (i.e., Illinois, Indiana, and Ohio), the GCM produces a less deleterious climate change scenario than does the RCM. In early spring, both models exhibit increases in precipitation but those of the GCM are larger (2–2.5 mm/day). In May, the GCM simulates mainly precipitation increases in Ohio and Indiana, whereas the RCM experiences decreases in all three states. In June and July, precipitation increases slightly for the GCM (~0.5 mm/day), but the RCM produces a complex pattern of increases and decreases. In August, both models experience decreases in precipitation but these decreases are intensified in the RCM case. Temperature increases in late spring and summer are about 1 °C higher for the RCM compared to those of the GCM, which remain between 4 °C and 5 °C, but increases tend to be smaller for the RCM in fall.

3.3. SOUTHEASTERN U.S. REGIONAL STUDY

We calculated the multi-year mean yields for the baseline climate case and for each of the six climate change cases for the southeastern study region for dryland maize and winter wheat, and for irrigated rice. Percent changes from the base yield obtained under each of the six climate change cases were also calculated. We examined the response of the CERES crop models on three different spatial scales: at the scale of the entire Southeast region (Tables VI–VIII), at the scale of yield aggregation for input to the Agricultural Sector Model, ASM (i.e., state level (Tables VI–VIII)), and at the 50 km grid scale (i.e., the scale of the observational and RCM data sets (Figures 6–8)).

Table VI

Percent changes in simulated yields of dryland corn from base for the coarse- and fine-resolution cases

State	Base yield (kg/ha)	% Change from base yield					
		GCM 330	RCM 330	GCM 540	RCM 540	GCM 540 + A	RCM 540 + A
Entire SE	*8100*	*–15* [a]	*–16*	*–2* [a, b]	*–2*	*6* [a, b]	*6* [b]
Alabama	7600	–12	–5	3 [b]	9	13	20
Arkansas	7400	–12	–5	2 [b]	9	22 [a]	25
Florida	6300	–21	–7 [b]	–5 [b]	10	–3 [b]	11
Georgia	7500	–19 [a]	–16	–5 [a, b]	–1 [b]	1 [a, b]	6 [b]
Louisiana	7900	–14	–9	–2 [b]	3 [b]	6 [a]	4 [b]
Mississippi	8400	–14	–8	–3 [b]	3 [b]	9 [a]	12
North Carolina	9100	–15	–34	–1 [b]	–18	–1 [b]	–9
South Carolina	8300	–19	–29	–5 [b]	–14	0 [a, b]	–6 [b]
Tennessee	9800	–14	–25	–3 [b]	–12	–1 [b]	–5

[a] Fine- and coarse-scenario yields are NOT significantly different ($\alpha = 0.05$).
[b] Climate change yield is NOT significantly different from base yield ($\alpha = 0.05$).
330 = case with climate change only.
540 = case with climate change plus elevated CO_2.
540 + A = case with climate change plus elevated CO_2 and adaptation.

3.3.1. *Entire Southeast Study Area*

At the level of the entire Southeast study area for the *climate change only cases*, all crops show decreases in yields compared to the base yields (Tables VI–VIII). There are no significant differences between the GCM and RCM scenario yields for corn and wheat (Tables VI and VII), whereas RCM scenario yield decreases are significantly larger than the GCM scenario decreases for rice (Table VIII). For the *elevated CO_2 cases*, simulated yield decreases are smaller compared to the climate change only cases, but the change from base conditions remains negative for all crops/scenarios (Tables VI–VIII). With *adaptation* for the entire Southeast study area, percentage changes in yield for corn and rice are positive for both climate scenarios. Wheat, which showed the least change in yield with adaptation from the elevated CO_2 case, still has negative yield changes, with larger losses simulated with the GCM scenario compared to the RCM.

For climate change only and elevated CO_2 cases, growing season lengths decreased for all crops (Table IX). For corn, growing season lengths decreased by about two weeks, compared to the baseline. Decreases for wheat ranged between 8 and 13 days, and for rice between 16 and 18 days. With adaptation, decreases were smaller, about 5 days for the RCM and 11 days for the GCM cases for corn as well as for winter wheat (2 to 6 days) and rice (about 4 days).

Table VII

Percent changes in simulated yields of dryland winter wheat from base for the coarse- and fine-resolution cases

State	Base yield (kg/ha)	% Change from base yield [b]					
		GCM 330	RCM 330	GCM 540	RCM 540	GCM 540 + A	RCM 540 + A
Entire SE	*4500*	*–36* [a]	*–32*	*–26* [a]	*–21*	*–25* [a]	*–21*
Alabama	4700	–35 [a]	–33	–24 [a]	–22	–24 [a]	–22
Arkansas	4400	–30	–24	–19	–12	–19	–12
Florida	3300	–65 [a]	–59	–59 [a]	–51	–58 [a]	–50
Georgia	4400	–37 [a]	–34	–26 [a]	–23	–25 [a]	–22
Louisiana	4000	–48	–38	–40	–28	–40	–28
Mississippi	4400	–38 [a]	–34	–28	–23	–28	–23
North Carolina	5200	–31 [a]	–29	–19 [a]	–18	–19 [a]	–17
South Carolina	4600	–30 [a]	–31	–18 [a]	–18	–18 [a]	–18
Tennessee	5100	–32 [a]	–27	–21 [a]	–16	–21 [a]	–15

[a] Fine- and coarse-scenario yields are NOT significantly different ($\alpha = 0.05$).
[b] Climate change yields are significantly different from base yield in all cases ($\alpha = 0.05$).
Definitions as in Table VI.

Table VIII

Percent changes in simulated yields of irrigated rice from base for the coarse- and fine-resolution cases

Region	Base yield (kg/ha)	% Change from base yield					
		GCM 330	RCM 330	GCM 540	RCM 540	GCM 540 + A	RCM 540 + A
Entire SE [a]	*9600*	*–16*	*–19*	*–3*	*–5*	*2*	*6*

[a] Data set included only 83 RCM grids. Climate change yields are significantly different from base yield in all scenarios. GCM and RCM yields are significantly different in all three comparisons ($\alpha = 0.05$).
Definitions as in Table VI.

Table IX

Duration of growing season (in days) for corn, winter wheat, and rice, for the entire Southeast study area, averaged over 36 growing seasons (35 growing seasons for winter wheat)

	Base	GCM 330	RCM 330	GCM 540	RCM 540	GCM 540 + A	RCM 540 + A
Corn	123	105	107	105	107	112	118
Winter wheat	214	206	201	206	203	208	212
Rice	128	110	112	110	112	122	126

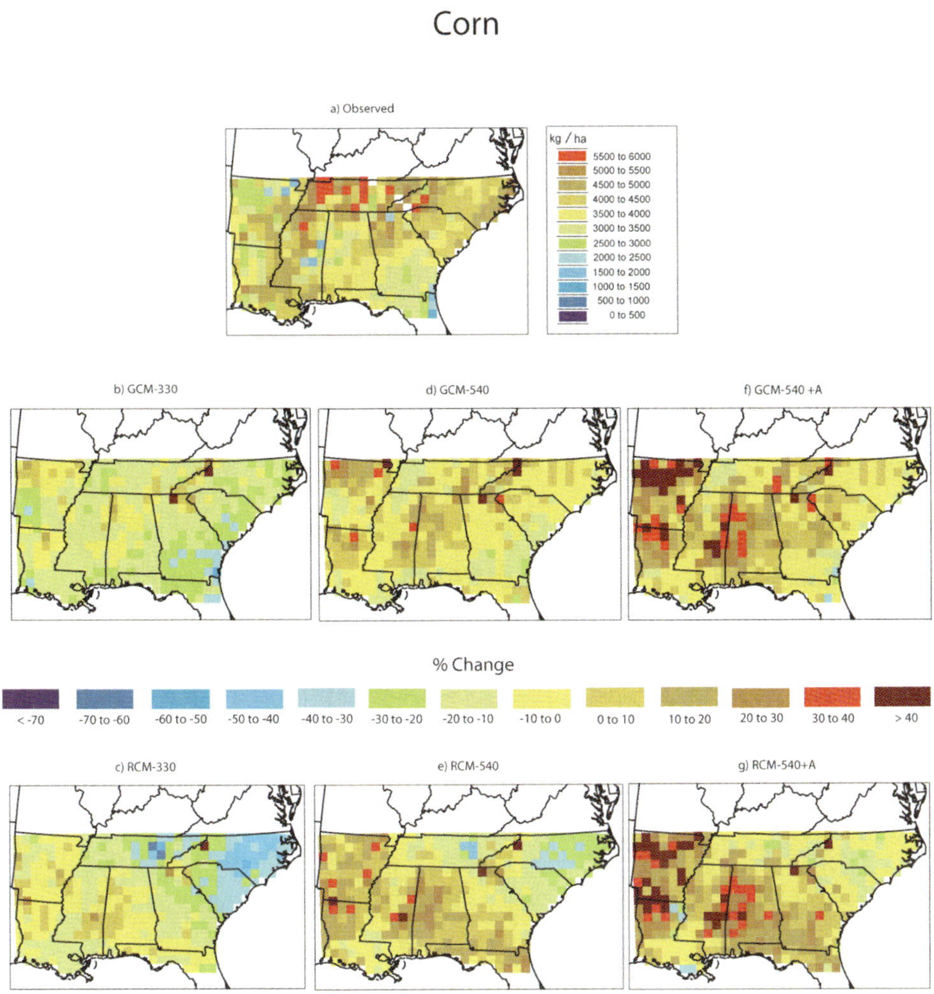

Figure 6. Simulated yields for dryland corn: (a) 36-year mean simulated baseline yields; (b) percent change in simulated yields under GCM climate change only case; (c) percent change in simulated yields under RCM climate change only case; (d) percent change in simulated yields under GCM elevated CO_2 case; (e) percent change in simulated yields under RCM elevated CO_2 case; (f) percent change in simulated yields under GCM adaptation case; (g) percent change in simulated yields under RCM adaptation case. All percent changes are calculated relative to the base case.

3.3.2. *State Level of Aggregation*

When we considered the changes in yield aggregated at the state level (the level of aggregation generally required by the economic model), contrasts were more complex. Corn yields, which had been similar for both scenarios on the scale of the entire Southeast study region, exhibit greater scenario contrasts on the state level (Table VI). For the climate change only cases, the GCM scenario produces significantly larger decreases for most states, compared to the RCM scenario, particularly

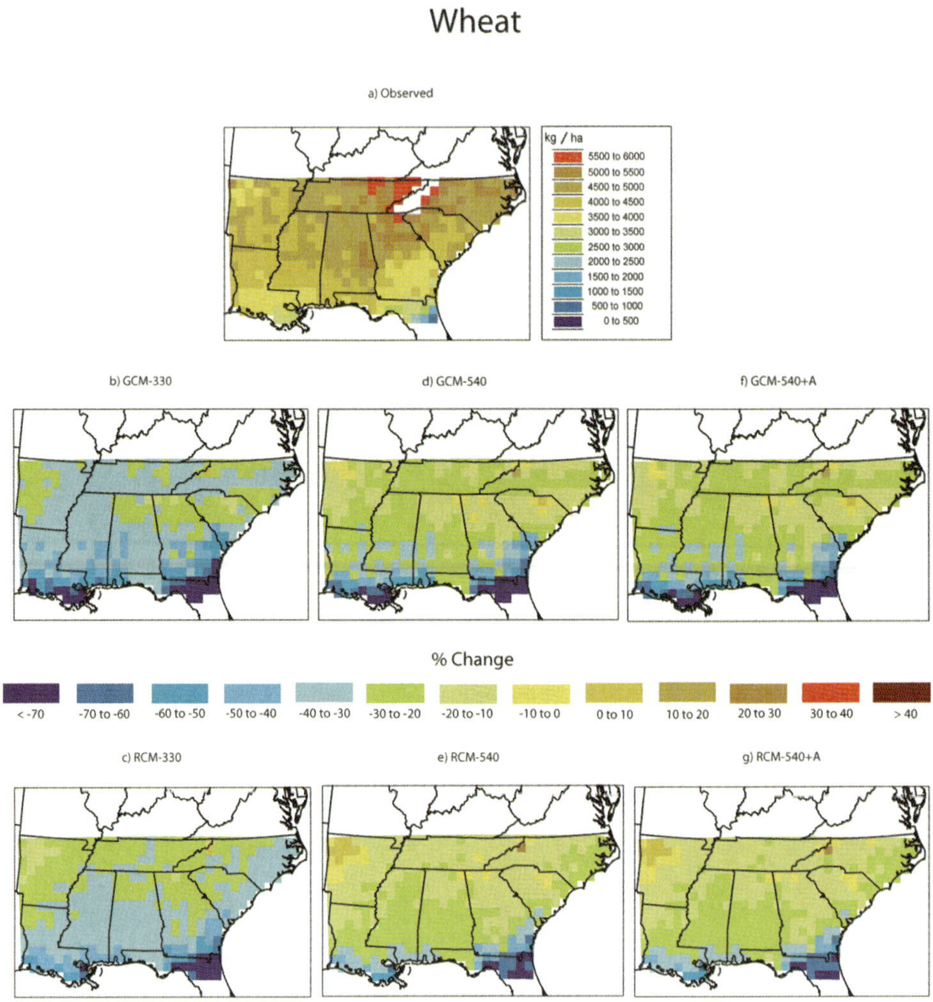

Figure 7. Simulated yields for dryland winter wheat: (a) 35-year mean simulated baseline yields; (b) percent change in simulated yields under GCM climate change only case; (c) percent change in simulated yields under RCM climate change only case; (d) percent change in simulated yields under GCM elevated CO_2 case; (e) percent change in simulated yields under RCM elevated CO_2 case; (f) percent change in simulated yields under GCM adaptation case; (g) percent change in simulated yields under RCM adaptation case. All percent changes are calculated relative to the base case.

in the southcentral and delta states. With elevated CO_2, corn yields showed percent increases in five states with the RCM scenario, but only in Alabama and Arkansas with the GCM scenario. The GCM yields in all states showed no significant difference from the baseline, while the RCM yields showed significant increases in three states and significant decreases in three states. This reflects the greater spatial variability in the RCM climate scenario. With adaptation, corn yields increased for most southern states for both scenarios, but still decreased for North and South

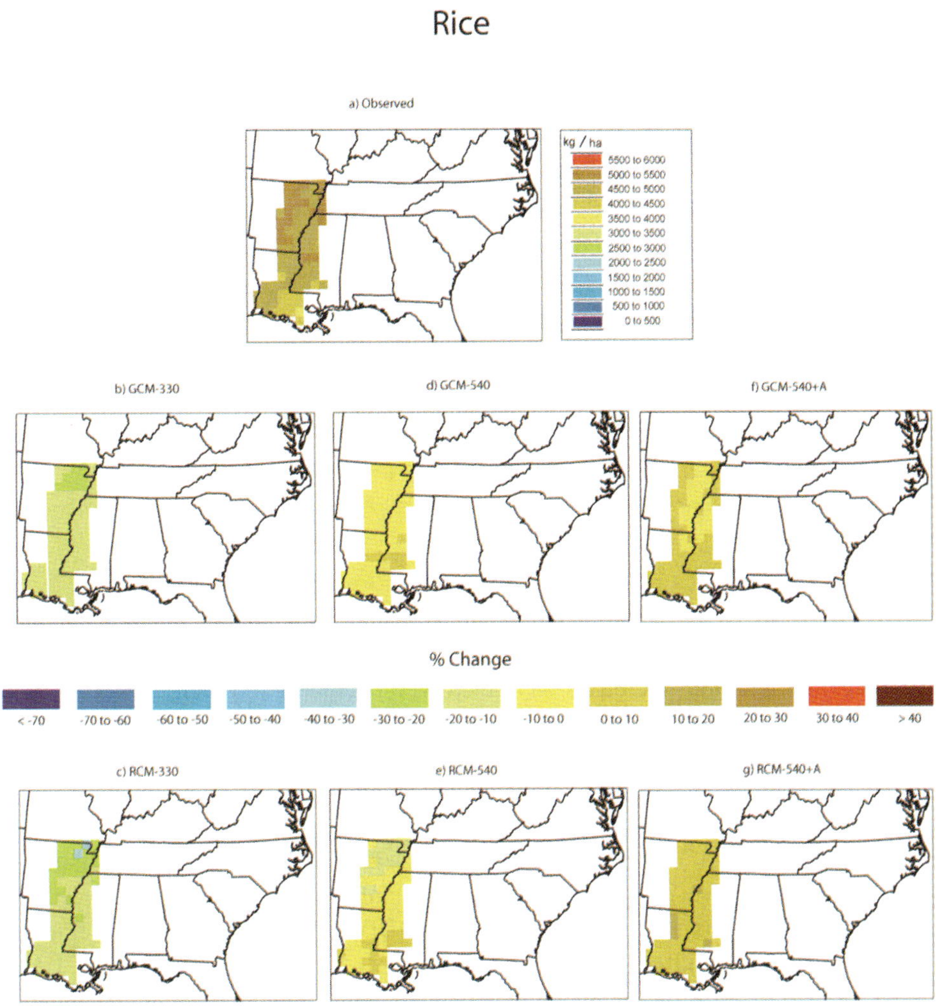

Figure 8. Simulated yields for irrigated rice: (a) 36-year mean simulated baseline yields; (b) percent change in simulated yields under GCM climate change only case; (c) percent change in simulated yields under RCM climate change only case; (d) percent change in simulated yields under GCM elevated CO_2 case; (e) percent change in simulated yields under RCM elevated CO_2 case; (f) percent change in simulated yields under GCM adaptation case; (g) percent change in simulated yields under RCM adaptation case. All percent changes are calculated relative to the base case.

Carolina and Tennessee for the RCM scenario. Contrasts in yield changes between the scenarios diminished under the adaptation case, indicating that adaptations mitigate the unique spatial scale effects of the climate change scenarios.

For corn, the temporal coefficient of variation (defined as the ratio of the standard deviation to the mean), a yield statistic used in the stochastic version of the ASM (Lambert et al., 1995), generally decreased for all the climate change cases compared to the baseline, especially in the elevated CO_2 and adaptation cases

Table X
Temporal coefficient of variation (CV) calculated over 36 seasons for dryland maize

State	Base	GCM 330	RCM 330	GCM 540	RCM 540	GCM 540 + A	RCM 540 + A
Alabama	0.24	0.22	0.20	0.19	0.17	0.17	0.15
Arkansas	0.27	0.21	0.20	0.18	0.16	0.11	0.11
Florida	0.31	0.27	0.21	0.23	0.18	0.26	0.20
Georgia	0.24	0.23	0.23	0.20	0.19	0.21	0.20
Louisiana	0.19	0.19	0.18	0.15	0.14	0.15	0.16
Mississippi	0.19	0.17	0.16	0.14	0.13	0.10	0.11
North Carolina	0.19	0.17	0.22	0.14	0.20	0.14	0.20
South Carolina	0.25	0.23	0.27	0.19	0.23	0.20	0.22
Tennessee	0.18	0.14	0.18	0.11	0.15	0.12	0.12

(Table X). This was true in all states except for North Carolina (all RCM cases) and South Carolina (RCM climate change only case), where substantial RCM-simulated temperature increases combined with decreases in precipitation resulted in the largest corn yield losses in the study region (Table VI). The GCM also simulated decreases in precipitation, but these were less extreme than in the RCM.

Wheat yields decrease considerably from the base for all cases for both GCM and RCM climate change scenarios. The differences between the yields calculated for the respective GCM and RCM cases, however, are generally not statistically significant (Table VII). This is due to both small yield differences between the scenarios in some states (the Carolinas, Table VII) and the effect of large temporal variability in yields in others (e.g., CV of over 0.5 in Florida). RCM and GCM winter wheat yield differences were statistically significant in only three states for the elevated CO_2 and adaptation cases, when the decreases in yield were larger for the GCM cases.

Under baseline conditions, temporal CVs of wheat were slightly greater in the south than in the north reflecting the marginality of climate of the southern states for winter wheat growth. Compared to the baseline, the temporal CV under climate change conditions (all cases) increased in all nine states, in some (Florida and Louisiana) more than doubling from baseline (e.g., CV of 0.54 for GCM and 0.48 for RCM elevated CO_2 cases compared to CV of 0.2 for baseline in Florida). Under climate change, temporal CVs of the yields simulated from the GCM were always greater than (or equal to) those of the yields from the RCM, and CVs were always higher for the southern tier of states compared to the northern tier, reflecting more frequent crop failures in the south and under the GCM (compared to the RCM) scenario. Under adaptation, temporal CVs were essentially the same as (or only slightly lower than) the CVs simulated for the climate change only and elevated

CO_2 cases. Because of the timing of the largest precipitation differences between the GCM and the RCM (i.e., mid to late summer, after the winter wheat harvest) RCM and GCM wheat yield differences in many states (especially in the north) were smaller than respective differences in corn yields. Corn yield differences developed partially in response to the large RCM and GCM precipitation contrasts during the second half of the growing season.

Under adaptation for corn (adjusting planting dates to planting earlier) we were able to mitigate the effect of decreased precipitation in the RCM scenario in the summer months and thus simulate overall yield increases, effectively minimizing the RCM and GCM yield contrasts. In the Southeast, where climate is already marginal for winter wheat growth, no planting date (and no cultivar) shifts were able to accommodate for the effect of increased winter temperatures and thus, even with adaptation, significant wheat yield losses were simulated. Note that one adaptation strategy that we tested was switching from winter wheat to spring wheat. However, since spring wheat yields are on average substantially lower than winter wheat yields, this crop switch still resulted in wheat yield losses over the Southeast. Yet with double cropping and appropriate rotation scheduling, this crop switch might be a viable strategy for mitigating the effects of a warmer climate in the study region.

Rice, an irrigated crop, was the only crop that showed statistically significant differences between the GCM and RCM scenarios at the level of the entire Southeast study region (the only level of analysis for rice) (Table VIII). Temporal variability in rice yields was also extremely low, rarely exceeding 0.1. Such an extremely low yield variability resulted in statistical significance of the seemingly small (i.e., 2%) yield differences (Table VIII).

3.3.3. *50-Km Grid Level of Aggregation*

On the 50-km grid level (Figures 6–8), yield changes were more spatially (and temporally) variable than on the state level. For corn, moderate yield decreases (10 to 30%) were simulated over most of the Southeast study region under both GCM and RCM climate change only conditions. Sporadic small yield increases were simulated in the western and central portions of the Southeast, for both GCM and RCM climate change only scenarios (Figure 6). The spatial pattern of corn yield changes was more complex and the range of change from baseline was greater for RCM compared to GCM (e.g., up to 60% decreases in yields in North Carolina and Tennessee for RCM). The spatial coefficients of variation (CVs), however, increased only slightly and were identical for the two scenarios (Table XI). For winter wheat, yield decreases from 10 to 70% were simulated under both the GCM and RCM scenarios (with larger yield losses in Florida and extreme southern Louisiana). The RCM scenario produced less severe yield decreases in the extreme south and north, compared to the GCM scenario (Figure 7). For both scenarios the spatial CV increased substantially over the baseline, more so for the GCM scenario (Table XI). In general, however, the lack of spatial contrast between the scenarios

Table XI

Spatial coefficient of variation (CV) for dryland maize and winter wheat, and for irrigated rice, for the southeastern United States

Crop	Base	GCM 330	RCM 330	GCM 540	RCM 540	GCM 540 + A	RCM 540 + A
Maize	0.21	0.22	0.22	0.19	0.18	0.16	0.15
Winter wheat	0.15	0.30	0.26	0.30	0.25	0.29	0.25
Rice	0.06	0.05	0.05	0.05	0.05	0.05	0.06

on the state level in wheat is also seen on the 50 km grid level. Rice exhibited an extremely homogeneous pattern of yield losses under the GCM climate change only case, mostly between 10% and 20%, with the largest losses in the north. Under the RCM climate change only case, only slightly greater yield losses were simulated in the northern portions of the domain (Figure 8). Rice yields were very homogeneous across the study area, with the spatial coefficient of variation between 0.05 and 0.06 across all seven cases (Table XI).

For all three crops, the positive effects of increased atmospheric CO_2 were seen for both relatively favorable environmental conditions (grid cells in the Mississippi river delta) and under conditions of environmental stress, such as the moisture deficit in North Carolina for the RCM scenario and extremely high temperatures in Florida. Under adaptation, conditions improve in most regions, but adaptations do not completely overcome the negative impacts of the climate change. Under adaptation, spatial CVs over the Southeast for corn and winter wheat were the lowest of all the climate change cases, for both RCM and GCM scenarios, and were even lower than in the base case for corn (Table XI). The spatial CVs, however, do not reveal a clear contrast in the effect of the RCM and GCM scenarios for any of the crops.

3.4. CAUSES OF CLIMATE CHANGE YIELD CONTRASTS

We investigated what aspects of the scenarios were most responsible for the contrasts in crop yields. Yield differences between the GCM and RCM climate change scenarios were regressed on climate variable differences during the growing season. We analyzed the best models containing one to three predictors. Table XII shows the best predictor models for most GCM grid cells for corn. Here we will only discuss grid cells for which the R^2 of the single variable model was greater than 0.30. We expect that our analysis will provide some general explanations for most parts of the Southeast study area, but not necessarily for all grid cells. Obviously the crop models integrate all aspects of the future climate in complex ways, and it is not always possible to explain this complexity via a regression

Table XII

Best models of corn yield differences using one, two, or three predictors, for selected GCM grids

% Change from base GCM RCM	CSIRO grid	Sample N	Number predictors	R^2	Variables in model (with sign of coefficients)
–5.6	SW	49	1	0.30	DP6 (+)
–1.5			2	0.48	DP6 (+), DSOL7 (+)
			3	0.58	DP6 (+), DSOL5 (+), DTMAX6 (–)
1.5	CW	71	1	0.58	DP6 (+)
8.5			2	0.68	DP6 (+), DP7 (+)
			3	0.73	DP6 (+), DP7 (+), DTMAX6 (–)
1.6	CC	60	1	0.58	DP5 (+)
–0.0			2	0.76	DP5 (+), DP6 (+)
			3	0.83	DP5 (+), DP6 (+), DTMIN5 (–)
–7.9	CE	35	1	0.74	DTMAX6 (–)
–19.9			2	0.84	DTMAX6 (–), DP5 (+)
			3	0.87	DP5 (+), DTMIN6 (–), DTMAX7 (–)
–0.9	NW	34	1	0.68	DTMIN6 (–)
–0.6			2	0.83	DTMAX5 (–), DSOL6 (+)
			3	0.89	DTMAX5 (–), DSOL6 (+), DP6 (+)

N = the number of 50 km grids analyzed.
DP = difference in precipitation (GCM–RCM).
DTMAX = difference in maximum temperature (GCM–RCM).
DTMIN = difference in minimum temperature (GCM–RCM).
DSOL = difference in solar radiation (GCM–RCM).
5 = May, 6 = June, 7 = July.

approach. However, Mearns et al. (2001a) did successfully explain the contrasts in climate changed yields based on the spatial scale of climate scenarios for the Erosion Productivity Impact Calculator (EPIC) crop model using this technique.

Statistical analysis discussed in this section was performed at the scale of a GCM grid cell (not at the state level or level of the entire Southeast study region, as in the previous sections). At the GCM grid level, we can perform statistical analysis of the effects of GCM and RCM climate changes on yields such that in one case (GCM) the climate change is constant throughout the entire GCM grid cell while in the other (RCM) the climate change varies throughout the GCM grid cell. This approach is consistent with that of Mearns et al. (2001a).

The contrast in the size of increase in precipitation in May and the contrast in direction of change in precipitation, from baseline, in June (particularly during the grain filling stage) were key explanatory factors for corn for many of the GCM grids, particularly for the central tier and the southwest grid (Table XII). Corn yield differences were positively related to the May and June precipitation differences (i.e., higher precipitation causes higher yields) (Figure 9 for the CC grid). In the central eastern (CE) grid of the Southeast (i.e., South Carolina), temperature contrast (GCM-RCM) in June was strongly negatively correlated with yields (i.e., higher temperature increases caused larger yield decreases). It provides the highest explained variance for this grid, and provides the highest single variable R^2 of all the grids. Temperature increases are up to 2 °C higher in the RCM scenario. Temperature controls the growing season length, individual growth stages, and actual dry matter production through the temperature stress term. The higher temperatures naturally also contribute to higher potential evapotranspiration rates. For the northwest grid, covering western Tennessee and northeastern Arkansas (Figure 1), minimum temperature in June provides the highest explanatory power, with the same relationship as with June maximum temperature for the CE grid. June minimum temperature increases from 0.7 °C to 1.7 °C, more in the GCM compared to the RCM in this area.

For winter wheat, temperature differences in February and March explain up to 90% of the variance in GCM and RCM yield differences in the southern and central tier of GCM grids. In the northern tier, the key explanatory variables also included April temperature differences. The GCM scenario yields were always lower than the RCM yields, corresponding with the fact that winter and spring temperature increases were generally higher for the GCM scenario. The winter temperature increase greatly affected vernalization, leading to widespread crop failures, more pronounced in the GCM scenario cases (e.g., 1027 incidents of crop failure in GCM versus 636 cases in RCM, for the elevated CO_2 cases over 35 seasons in the entire Southeast study region). More frequent crop failures in the GCM cases in turn lead to substantially lower mean yields under the GCM compared to RCM scenario (for example in the southwest grid, GCM yields decreased by 41% but RCM yields only by 30%). Specifically, crop failures were a result of the crop failing to vernalize completely, and thus the crop does not reach the point of terminal spikelet formation. Hence, grain fill is never reached and no yield is produced. Frequent crop failures in the climate change cases translated into substantially larger temporal yield variability compared to the baseline (see Section 3.3).

For rice, the key explanatory variables are July maximum (and to a lesser degree, minimum) air temperature differences between GCM and RCM, which explain 67% of the variance. Since rice is irrigated, it is the temperature and not precipitation contrasts that drive rice yield. Maximum temperature increases are higher for the RCM than the GCM in the rice growing area of the Southeast for the

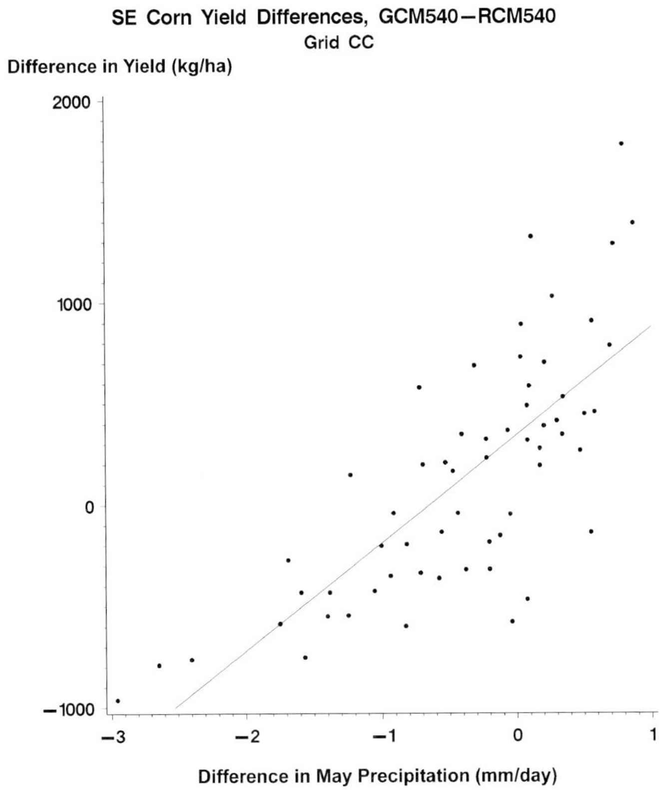

Figure 9. Scatter plot with linear regression line of difference in climate changed yields (GCM-RCM) versus the difference in the climate changed climate variable (GCM-RCM) for May precipitation (mm/day) for the CC grid. Difference in yield of dryland corn is in kg/ha.

climate change only and elevated CO_2 cases, and RCM rice yields tend to be lower than GCM yields. The differences in rice yields (GCM versus RCM), although statistically significant, are quite small in absolute values.

3.5. CONTERMINOUS UNITED STATES (EXCLUDING THE SOUTHEAST)

While our research focused on the southeastern United States, yield changes in the rest of the country, under different climate change scenarios, were required by the economic model, ASM (see Adams et al., 2003, for a discussion of economic model results). The ASM was only applied to the elevated CO_2 and adaptation cases. Yield changes over the conterminous United States are also discussed in Mearns et al. (2003a). For both the elevated CO_2 and adaptation cases for dryland corn, the RCM scenario produced greater yield increases in the Great Plains compared to the GCM (e.g., for the elevated CO_2 case RCM simulated yield increases of up to 50% and GCM yield increases of up to 30%). Yield decreases over the eastern states were greater for the RCM compared to the GCM scenario. Over most of the Midwest (with the exception of the upper Midwest, i.e., Minnesota and Wisconsin, where the RCM scenario resulted in yield increases), RCM scenario produced greater maize yield decreases compared to GCM.

For winter wheat for elevated CO_2 cases, both scenarios resulted in yield increases in the western mountain states, the northern Great Plains, and the upper Midwest. While the GCM produced higher yields in Idaho (e.g., yield increases by 60% for the GCM and 13% for the RCM), North Dakota, Oregon, Utah, and Virginia, the RCM generated greater yield increases compared to the GCM in Michigan, Minnesota, South Dakota, and New Jersey. In the lower Midwest and the southern Great Plains, both scenarios resulted in yield losses, with the GCM generally resulting in greater yield losses compared to the RCM (with the exception of Iowa, where RCM yields were lower). Under adaptation for winter wheat, while the yields uniformly went up, the relative yield changes between the GCM and RCM climate change scenarios were the same as in the elevated CO_2 cases.

For rice in California and eastern Texas, the RCM scenario resulted in smaller yield decreases (e.g., 18% and 11% decreases in California and Texas, respectively) compared to the GCM (25% and 13% decrease) for the elevated CO_2 case. In contrast, in Missouri the RCM produced greater yield losses compared to the GCM for the elevated CO_2 case (i.e., 14% versus 6% losses). In the coastal plains of Texas, both scenarios produced essentially no yield change from base under elevated CO_2 conditions. Under adaptations in California, under both climate change scenarios rice yields increased, more so for the GCM (i.e., 14% increase for the GCM and 8% increase for the RCM). In contrast, in Texas (both eastern and coastal plains regions), the RCM resulted in greater yield increases under adaptation compared to the GCM.

4. Discussion and Conclusions

Our analysis indicates that the spatial scale of climate change scenarios can substantially affect the simulation of changes in crop yields on several levels of spatial aggregation, but that the effect is also crop specific. Under dryland conditions, corn was the most affected and wheat the least. The effect is especially apparent at the finest spatial resolution (i.e., individual 50 km grid boxes), but persists up to the state level. The contrasts are averaged out on the domain-wide level, however. We also determined the differences in main variables of the two climate change scenarios that were most responsible for the contrasts in the climate changed yields. The most important explanatory variables varied with location and crop. For corn the degree of precipitation increase in May and direction of change in precipitation in June accounted for most of the contrast in climate changed yields. Temperature variables were most important for wheat and rice.

A logical question to ask at the end of this study is: What is the added value of finely spatially distributed climatic change information? Our results suggest that the answer to this question depends on the spatial unit of interest (individual grid cells versus regional impacts), as well as on the crop studied and the number of environmental and management factors accounted for in the analysis. We see largest differences between the crop yield effects of the RCM and GCM scenarios when climate change only conditions are imposed. However, as we account for additional factors, such as physiological effects of elevated CO_2 levels and effects of adaptations, the unique spatial scale effects of the climate change scenarios tend to diminish, but are not eliminated.

Comparing the effects of the two scenarios for the Southeast on the state level, we find that the RCM scenario is less deleterious (or more beneficial) for corn in the west and central portions of the study area but more deleterious (less beneficial) in the Carolinas and Tennessee than the GCM scenario. In the case of wheat, the RCM scenario is slightly less damaging than the GCM scenario throughout the study area. This is not necessarily the case for the other crops included in the Southeast study (Carbone et al., 2003; Doherty et al., 2003).

It should be noted that we cannot at this point demonstrate that one scenario is in reality any more likely than the other. While the inclusion of spatial detail in scenarios clearly affects results, we cannot definitively conclude that we have more confidence in the results with higher resolution. Evidence is mounting, however, that in regions with complex topography, coastlines, or land use, regional climate models with higher resolution may very well respond more realistically to external forcings such as from increased concentrations of greenhouse gases, than coarse resolution GCMs (Giorgi et al., 2001). Moreover it has now been demonstrated numerous times that simulations with regional models tend to better reproduce the current climatology of many regions than similar simulations with coarse models. Such is the case for the simulations used for the scenarios applied here (see Mearns et al., 2003b).

This study is fundamentally a sensitivity analysis of the scale effect of scenarios, and we have demonstrated that the impacts results for some crops at certain levels of aggregation are sensitive to the spatial scale of scenarios. Elsewhere in this issue Adams et al. goes on to demonstrate that with further complex aggregation of cropping results and application to an agricultural economic model, the original spatial scale effect of the scenarios is still apparent.

We will have to wait for more research before being able to establish definitively the added value of high resolution climate change information. For this to come about the uncertainty of the spatial scale of the climate change information will also have to be appropriately situated in the context of the myriad other uncertainties regarding the prediction of climate change, such as the paths of future greenhouse gas emissions and the uncertainties concerning global climate model sensitivity.

Another remaining uncertainty is based on the particular crop model used to estimate the impact of changes in climate (e.g., Mearns et al., 1999). This is particularly problematic since it is usually not possible to select the 'best' crop model to use based solely on validation results, any more than it is possible to

IPCC: 1996, 'Climate Change 1995: The Science of Climate Change, Contribution of Working Group I to the Second Assessment Report of the Intergovernmental Panel on Climate Change', in Houghton, J. T., Meira Filho, L.G., Callander, B. A., Harris, N., Kattenberg, A., and Maskell K. (eds.), Cambridge University Press, Cambridge, U.K., pp. 572.

Jones C. A. and Kiniry, J. R. (eds.): 1986, 'CERES-Maize: A Simulation Model of Maize Growth and Development', *Texas A&M University Press*, 194 pp.

Kiniry J. R. and Bockholt, A. J.: 1998, 'Maize and Sorghum Simulation in Diverse Texas Environments', *Agron. J.* **90**, 682–687.

Kiniry J. R., Williams, J. R., Vanderlip, R. L., Atwood, J. D., Reicosky, D. C., Mulliken, J., Cox, W. J., Mascagni, H. J., Hollinger, S. E., and Wiebold, W. J.: 1997, 'Evaluation of Two Maize Models for Nine U.S. Locations', *Agron. J.* **89**, 421–426.

Lambert D. K., McCarl, B. A., He, Q., Kaylen, M. S., Rosenthal, W., Chang, C. C., and Nayda, W. I.: 1995, 'Uncertain Yields in Sectoral Welfare Analysis: An Application to Global Warming', *J. Agr. Appl. Econ.* **27**, 423–436.

Littell R. C., Milliken, G. A., Stroup, W. W., and Wolfinger, R. D.: 1996, *SAS System for Mixed Models*, SAS Institute Inc., Cary, NC, U.S.A., 633 pp.

Liu W. T. H., Botner, D. M., and Sakamoto, C. M.: 1989, 'Application of CERES-Maize Model to Yield Prediction of a Brazilian Maize Hybrid', *Agric. For. Meteorol.* **45**, 299–312.

Mall R. K. and Aggarwal, P. K.: 2002, 'Climate Change and Rice Yields in Diverse Agro-environments of India. I. Evaluation of Impact Assessment Models', *Clim. Change* **52**, 315–330.

Mascagni, H. J. Jr., Kang, S. M., and Burns, D. R.: 1998, 'Influence of Planting Date and Hybrid Maturity on Corn Yield Performance and Plant Development on Sharkey Clay'.

McCarl B. A., Chang, C. C., Atwood, J. D., and Nayda, W. I.: 2000, 'The U.S. Agricultural Sector Model', *http://agecon.tamu.edu/faculty/mccarl/asm.html.*

McMaster G. S., Wilhelm, W. W., and Morgan, J. A.: 1992, 'Simulating Winter Wheat Shoot Apex Phenology', *J. Agricultural Science Cambridge* **119**, 1–12.

Mearns L. O., Carbone, G., Doherty, R. M., Tsvetsinskaya, E. A., McCarl, B. A., Adams, R. M., and McDaniel, L.: 2003a, 'The Uncertainty Due to Spatial Scale of Climate Scenarios in Integrated Assessments: An Example from U.S. Agriculture', *Integrated Assessment*, accepted.

Mearns, L. O., Easterling, W., Hays, C., and Marx, D.: 2001a, 'Comparison of Agricultural Impacts of Climate Change Calculated from High and Low Resolution Climate Model Scenarios: Part I. The Uncertainty of Spatial Scale', *Clim. Change* **51**, 131–172.

Mearns L. O., Giorgi, F., Shields, C., and McDaniel, L.: 2003b, 'Climate Scenarios for the Southeastern U.S. Based on GCM and Regional Model Simulations', *Clim. Change* **60**, 7–35.

Mearns, L. O., Hulme, M., Carter, T. R., Leemans, R., Lal M., and Whetton, P.: 2001b, 'Climate Scenario Development', Chapter 13 in Houghton et al., *IPCC Third Assessment Report. The Science of Climate Change*, Cambridge University Press: Cambridge, pp. 739–768.

Mearns L. O., Mavromatis, T., Tsvetsinskaya, E., Hays, C., and Easterling, W.E.: 1999, 'Comparative Responses of EPIC and CERES Crop Models to High and Low Spatial Resolution Climate Change Scenarios', *J. Geophys. Res.* **104**, 6623–6646.

Mearns L. O., Rosenzweig, C., and Goldberg, R.: 1992, 'Effect of Changes in Interannual Climatic Variability on CERES-Wheat Yields: Sensitivity and 2 × CO_2 General Circulation Model Studies', *Agric. For. Meteorol.* **62**, 159–189.

Mearns L. O., Rosenzweig, C., and Goldberg, R.: 1996, 'The Effect of Changes in Daily and Interannual Climatic Variability on CERES-Wheat: A Sensitivity Study', *Clim. Change* **32**, 257–292.

Mearns L. O., Rosenzweig, C., and Goldberg, R.: 1997, 'Mean and Variance Change in Climate Scenarios: Methods, Agricultural Applications, and Measures of Uncertainty', *Clim. Change* **35**, 367–396.

Moulin, A. P. and Beckie, H. J.: 1993, 'Evaluation of the CERES and EPIC Models for Predicting Spring Wheat Grain Yield over Time', *Can. J. Plant Sci.* **73**, 713–719.

Pang X. P., Letey, J., and Wu, L.: 1997, 'Yield and Nitrogen Uptake Prediction by CERES-Maize Model Under Semiarid Conditions', *Soil Sci. Soc. Am. J.* **61**, 254–256.

Parry M., Rosenzweig, C., Iglesias, A., Fischer, G., and Livermore, M.: 1999, 'Climate Change and World Food Security: A New Assessment', *Global Environ. Change* **9**, 51–67.

Phillips J. G., Cane, M. A., and Rosenzweig, C.: 1998, 'ENSO, Seasonal Rainfall Patterns and Simulated Maize Yield Variability in Zimbabwe', *Agric. For. Meteorol.* **90**, 39–50.

Reilly, J. et al.: 2003, 'U.S. Agriculture and Climate Change: New Results', *Clim. Change* **57**, 43–69.

Ritchie J. T.: 1985, 'A User Oriented Model of the Soil Water Balance in Wheat', in Day W. and Atykin, R. K. (eds.), *Wheat Growth and Modeling*, Plenum Press.

Ritchie J. T., Alocilja, E. C., Singh, U., and Uehara, G.: 1986, 'IBSNAT and the CERES-Rice Model', in *Weather and Rice*, Published by IRRI.

Ritchie J. T. and Otter, S.: 1985, 'Description and Performance of CERES-Wheat: A User-Oriented Wheat Yield Model', Chapter 10 in Willis, W. O. (ed.) *ARS Wheat Yield Project*, United States Department of Agriculture, Agricultural Research Service, pp. 159–175.

Rosenzweig C.: 1990, 'Crop Response to Climate Change in the Southern Great Plains: A Simulation Study', *Professional Geographer* **42** (1), 20–37.

Rosenzweig C. and Parry, M. L.: 1994, 'Potential Impact of Climate Change on World Food Supply', *Nature* **367**, 133–138.

Southworth J., Pfeifer, R. A., Habeck, M., Randolph, J. C., Doering, O. C., and Rao, D. G.: 2002, 'The Sensitivity of Winter Wheat Yields in the Midwestern United States to Future Changes in Climate, Climate Variability, and CO_2 Fertilization', *Clim. Res.* **22**, 73–86.

Southworth J., Randolph, J. C., Habeck, M., Doering, O. C., Pfeifer, R. A., Rao, D. G., and Johnston, J. J.: 2000, 'Consequences of Future Climate Change and Changing Climate Variability on Maize Yields in the Midwestern United States', *Agriculture, Ecosystems and Environment* **82**, 139–158.

Thomson, A. M., Brown, R. A., Ghan, S. J., Izaurralde, R. C., Rosenberg, N. J., and Leung, L. R.: 2002, 'Elevation Dependence of Winter Wheat Production in Eastern Washington State With Climate Change: A Methodological Study', *Clim. Change* **54** (1), 141–164.

Tubiello F. N., Rosenzweig, C., Goldberg, R. A., Jagtap, S., and Jones, J. W.: 2002, 'Effects of Climate Change on U.S. Crop Production: Simulation Results Using Two Different GCM Scenarios. Part I: Wheat, Potato, Maize, and Citrus', *Clim. Res.* **20** (3), 259–270.

Tubiello F. N., Rosenzweig, C., Kimball, B. A., Pinter, P. J., Wall, G. W., Hunsaker, D. J., LaMorte, R. L., and Garcia, R. L.: 1999, 'Testing CERES-Wheat With Free-Air Carbon Dioxide Enrichment (FACE) Experiment Data: CO_2 and Water Interactions', *Agron. J.* **91** (2), 247–255.

USDA: 1994, 'State Soil Geographic (STATSGO) Data Base Data Use Information', U.S. Dept. of Agric., Soil Conser. Serv., National Soil Survey Center. Misc. Publ. No. 1392.

Wu Y., Sakamoto, C. M., and Botner, D. M.: 1989, 'On the Application of the CERES-Maize Model to the North China Plain', *Agric. For. Meteorol.* **49**, 9–22.

(Received 12 July 2002; in revised form 13 March 2003)

RESPONSE OF SOYBEAN AND SORGHUM TO VARYING SPATIAL SCALES OF CLIMATE CHANGE SCENARIOS IN THE SOUTHEASTERN UNITED STATES

GREGORY J. CARBONE [1], WILLIAM KIECHLE [1], CHRISTOPHER LOCKE [1], LINDA O. MEARNS [2], LARRY McDANIEL [2] and MARY W. DOWNTON [2]

[1] *Department of Geography, University of South Carolina, Columbia, SC 29208, U.S.A.*
E-mail: greg.carbone@sc.edu
[2] *National Center for Atmospheric Research*

Abstract. This study examines how uncertainty associated with the spatial scale of climate change scenarios influences estimates of soybean and sorghum yield response in the southeastern United States. We investigated response using coarse (300-km, CSIRO) and fine (50-km, RCM) scale climate change scenarios and considering climate changes alone, climate changes with CO_2 fertilization, and climate changes with CO_2 fertilization and adaptation. Relative to yields simulated under a current, control climate scenario, domain-wide soybean yield decreased by 49% with the coarse-scale climate change scenario alone, and by 26% with consideration for CO_2 fertilization. By contrast, the fine-scale climate change scenario generally exhibited higher temperatures and lower precipitation in the summer months resulting in greater yield decreases (69% for climate change alone and 54% with CO_2 fertilization). Changing planting date and shifting cultivars mitigated impacts, but yield still decreased by 8% and 18% respectively for the coarse and fine climate change scenarios. The results were similar for sorghum. Yield decreased by 51%, 42%, and 15% in response to fine-scale climate change alone, CO_2 fertilization, and adaptation cases, respectively – significantly worse than with the coarse-scale (CSIRO) scenarios. Adaptation strategies tempered the impacts of moisture and temperature stress during pod-fill and grain-fill periods and also differed with respect to the scale of the climate change scenario.

1. Introduction

Increasing concentrations of greenhouse gases in the atmosphere will likely alter global temperature and precipitation patterns during the next century (Houghton et al., 2001). The agricultural impacts of such changes would depend on the magnitude and spatial expression of change, and farmers' ability to adapt. Agriculture in the southeastern United States could be particularly vulnerable to climatic change: current summer maximum temperatures in the region often exceed 32 °C, evaporation rates exceed growing season precipitation, and agricultural soils typically have relatively low water-holding capacity and low fertility. Moreover, the profitability of farming in the Southeast depends on relatively low capital inputs, thus limiting some options for mitigating the impacts of climate change.

Some of the uncertainty surrounding estimates of regional agricultural response to future climate change comes from the tools used to measure crop response to environmental change. The influence of climate change on crop yield has been

Climatic Change **60:** 73–98, 2003.

estimated with a variety of crop models that differ with respect to complexity and the degree to which they incorporate physiological processes (Boote et al., 1996). Some studies have demonstrated that different crop models produce different yield response to the same climate conditions (Xie et al., 2001; Wolf, 2002a). Others have examined the influence of varying soil type (Wassenaar et al., 1999), carbon dioxide level (Brown et al., 2000; Haskett et al., 2000), or field management, including adaptation strategies related to planting date, cultivar, fertilization, and irrigation (Curry et al., 1990; Rosenzweig and Iglesias, 1998; Alexandrov and Hoogenboom, 2000; Tubiello et al., 2000; Wolf, 2002b).

Climate change scenarios and the means by which they are incorporated into crop simulation models present another uncertainty in agricultural impact studies. Researchers generally recognize such uncertainties and adjust for them in a variety of ways. For example, it is widely recognized that GCMs were not designed to make projections at a regional scale. Therefore, many impact studies have used output from more than one GCM to create a range of possible scenarios, expressing the uncertainty associated with regional-scale projections (e.g., Curry et al., 1995; Alexandrov and Hoogenboom, 2000). It is also clear that GCM grids are coarse relative to the scale for which crop simulation models were developed. Researchers have accommodated these scale differences by downscaling GCM output to individual simulation sites. The latter approach has been accomplished statistically (Semenov and Barrow, 1997) and by interpolation from the GCM grid points to smooth changes spatially (Alexandrov and Hoogenboom, 2000). Some have created a physically-based climate change scenario by nesting relatively finer scale regional climate models within a GCM (Easterling et al., 2001; Guerena et al., 2001; Mearns et al., 2001). Preliminary evidence from these studies suggests that the spatial scale of the climate change scenario influences the magnitude and spatial patterns of agricultural impacts. Here we investigate that premise further.

We examine the potential impacts of climate change on soybean and sorghum yield in the Southeast United States using two scenarios of climate change that differ in spatial resolution. A coarse scenario is produced from GCM output, while a fine scenario is produced by nesting a regional climate model within the GCM. This allows us to compare crop response relative to the scale at which we express climate change. We consider these differences with respect to crop development and stress, and examine how a wide range of adjustments could mitigate yield decreases resulting from climate change. Section 2 discusses the crop models and climate change scenarios used, and outlines the simulation experiments. We present and interpret our results in Section 3 and make summary remarks in Section 4.

2. Methods and Data

2.1. CROP MODELS

We simulated soybean and sorghum yield in 414 50-km grid cells across the Southeast using the CROPGRO-Soybean and CERES-Sorghum v. 3.1 crop simulation models which are part of the International Benchmark Sites Network for Agrotechnology Transfer (IBSNAT) suite of crop simulation tools (Tsuji et al., 1994). These models simulate carbon, water, and nitrogen balances for the plant and soil (Hoogenboom et al., 1994; Boote et al., 1998). They have been used extensively in climate impact studies because their structure links plant processes to environmental and management inputs. CROPGRO-Soybean has been validated across a range of different environmental conditions and has shown skill in predicting phenological stages, biomass accumulation, and yield (Brisson et al., 1989; Colson et al., 1995; Kiniry et al., 1997; Boote et al., 1997; Carbone et al., 2003). CERES-Sorghum has also been calibrated and tested in diverse environments (Birch et al., 1990; Robertson et al., 1993; Hammer and Muchow, 1994; Castrignano et al., 1997).

While physiologically based, the IBSNAT crop simulation models require inputs that are either readily available or can be adequately estimated. These include meteorological inputs such as daily solar radiation, maximum and minimum temperatures, and precipitation. We acquired daily maximum and minimum temperatures, and precipitation from the National Weather Service cooperative network. Solar radiation was estimated stochastically using the synthetic weather generator, WXGEN (Richardson and Nicks, 1990). Mearns et al. (2003a, this issue) provide further details on the creation of the baseline climate data set. The crop models also require the following soils information: lower limit of plant-extractable soil water, drained upper limit soil water content, and saturated water content for each soil layer. We estimated values for these variables empirically, using a regression model relating soil texture characteristics to field capacity and wilting point (Baumer et al., 1994). Soil texture properties from one soil in each 50-km grid was selected from the State Soil Geographic (STATSGO) data base (Reybold and TeSelle, 1989). We considered only ‘prime farmland’ soils and selected the agricultural soil type that occupied the greatest area of each grid. While the soils differ between grid cells, some general spatial patterns exist across the region. Sands and loamy sands cover most of the Atlantic Coastal Plain, sandy loams dominate areas further inland (in the Piedmont), and silt loams cover most of the western portion of the study region, including the Mississippi Delta. The models also need field management inputs, including cultivar, sowing date, plant population, row spacing, sowing depth, and irrigation scheduling. A series of cultivar coefficients express the duration of growth stages, maximum growth rates, photoperiod, threshold temperatures, and grain or seed characteristics (Table I).

Table I

Soybean genotype coefficients for CROPGRO-Soybean simulations

Maturity	CSDL	PPSEN	EM–FL	FL–SH	FL–SD	SD–PM	FL–LF
I	13.84	0.203	17.0	6.0	13.0	32.0	26.0
IV	13.09	0.294	19.4	7.0	15.0	34.5	26.0
V	12.83	0.303	19.8	8.0	15.5	35.0	18.0
VII	12.33	0.320	20.8	10.0	16.0	36.0	18.0

CSDL: Critical short day length below which reproductive development progresses with no daylength effect (hours).
PPSEN: Slope of the relative response of development to photoperiod with time ($hour^{-1}$).
EM-FL: Time between plant emergence and flower appearance (photothermal days).
FL-SH: Time between first flower and first pod (photothermal days).
FL-SD: Time between first flower and first seed (photothermal days).
SD-PM: Time between first seed and physiological maturity (photothermal days).
FL-LF: Time between first flower and end of leaf expansion (photothermal days).

In the initial baseline and climate change crop simulations, we used one uniform planting date (calendar day 150 – approximately May 30) and cultivar across the entire region. We used published values for soybean coefficients from the generic maturity group V and sorghum coefficients from Cargill 837 (Tsuji et al., 1994). While planting date and variety vary over the study period and region, this strategy allowed us to control for these management variables and focus on the impacts of climate change. We consulted National Agricultural Statistics Service (USDA, 1997) and state agricultural extension reports, and county agricultural extension agents to select one planting date and cultivar without straying far from historic and current management practices. The impact of this strategy was minor as discussed in the adaptation results below. For all simulations, the CROPGRO model was run with water balance, soil-N balance, and plant-N balance options turned on. The Priestley-Taylor method (Priestley and Taylor, 1972) was used to compute crop potential evapotranspiration. The model was run in 'sequence mode' through consecutive years in order to preserve soil moisture values from one season to the next. We simulated yield in every grid cell in the entire region. While agricultural statistics show variability in production across the Southeast, every southeastern state produces at least some soybean and sorghum, and climatic change could allow production in areas not currently used.

We simulated baseline soybean and sorghum yields for the 36-year period, 1960–1995 using observed daily meteorological data from the National Weather Service cooperative network. One station was selected to represent each of the regional climate model 50-km grids on the basis of proximity to the grid center and data quality (see Mearns et al., 2003a). Then we reran the models for three cases: (1) climate change only; (2) climate change plus direct CO_2 fertilization effects –

referred hereafter as the 'elevated-CO_2 case'; and (3) the elevated-CO_2 case with adaptation. We used an ambient CO_2 value of 330 ppm for the baseline and climate change only cases, and 540 ppm for the elevated-CO_2 cases. CROPGRO-Soybean and CERES-Sorghum consider the effects of CO_2 fertilization by adjusting photosynthesis and leaf stomatal resistance in response to ambient CO_2 concentrations (Boote and Pickering, 1994; Pickering et al., 1995). Pickering et al. (1995) reported modeled photosynthesis increases of 30% and potential evapotranspiration decreases of 5% in response to a doubling of ambient CO_2. We performed simulations for control and climate change scenarios using both dryland and irrigated options. While we will report primarily on dryland simulations, the irrigated runs provide a standard to investigate the influence of water stress.

2.2. CLIMATE CHANGE SCENARIOS

Climate change scenarios for the southeastern United States were constructed at two different spatial resolutions. Controlled and doubled CO_2 equilibrium experiments of the Australian CSIRO Mk2 GCM (Watterson, 1998; Watterson et al., 1999) were used to generate a coarse resolution (approximately 3.2° latitude × 5.6° longitude) scenario with 11 grid boxes across the Southeast. A nested regional climate model (RCM), driven by the CSIRO boundary conditions, was used to create a fine resolution (50-km) scenario for 414 grid boxes across the region. The general features of the two scenarios are similar, but important differences often exist at the subregional scale (Giorgi et al., 1998; Mearns et al., 2003a). For example, the CSIRO model predicts greatest June, August, and September temperature increases in the western portion of the study region, while June–September RCM temperature increases are highest near the Atlantic. While both models simulate increases in summer temperature, the RCM increases are greater. CSIRO simulates increases in June precipitation for most of the Southeast, whereas RCM simulates mainly decreases. Both models simulate precipitation decreases during July, August, and September, but the decreases are more severe for the RCM, especially along the Atlantic Coastal Plain and Piedmont. Temperature and precipitation changes from RCM are more spatially variable than those produced by CSIRO. Mearns et al. (2003a) provide further details on the climate change scenarios.

Output from CSIRO and RCM, representing climate change with CO_2 doubling, was used to adjust the maximum and minimum temperature, precipitation, and solar radiation of a 36-year (1960–1995) observed climate data set. Observed data from one station within each 50-km RCM grid were used. In the coarse-resolution scenario, monthly mean temperature differences ($2 \times CO_2$ – baseline) or precipitation ratios ($2 \times CO_2$/baseline) from the CSIRO model were imposed uniformly on all 50 km grids encompassed within each CSIRO grid. The high-resolution scenario was created by imposing the regional model changes uniquely on observations in each 50-km grid.

To provide input for an agricultural economic model of the U.S., simulations were also run for stations elsewhere in the United States to account for simultaneous climate change impacts outside the Southeast. We used observed data from selected stations from the period 1961–1985 to simulate baseline yield in other important soybean- and sorghum-growing regions. Climate change scenarios were created for these other stations using the same GCM results and additional runs with the same RCM in other areas of the U.S. We measured the impacts of climate change at these other stations by comparing baseline yield against the two climate change scenarios. Yield changes for these other regions are also discussed in Mearns et al. (2003b). The results were incorporated into the Agricultural Sector Model (ASM; see Adams et al., 2003).

2.3. ADAPTATION CASES

In an additional series of simulations, we adjusted planting date and cultivar characteristics – two adaptation strategies that growers could use to mitigate the impacts of climate change. For soybean, we simulated yield using four different maturity groups (I, IV, V, and VII), each with eight planting dates (calendar days 100, 110, 120, 130, 140, 150, 160, and 170). These adjustments were designed to alter growing season length and change the timing of phenological events relative to the most stressful part of the growing season. Table I outlines how crop coefficients vary among the four cultivars. For sorghum, we adjusted crop growth parameters to shorten the juvenile and grain-filling stages, and moved the planting date from calendar day 150 to calendar day 100 to avoid the hottest part of the summer. For both crops, we determined the combination of cultivar and planting date that produced the highest 36-year average yield at both the grid cell and state-wide level and used this yield value to measure the impacts of climatic change with adaptation. Table II summarizes all cases.

2.4. MEASURING YIELD RESPONSE

We calculated the 36-year average yield for all 50-km grid cells for the three cases and two climate change scenarios. Percentage yield change (from the baseline simulation) for each was calculated as:

$$\Delta Y = \frac{Y - Y_b}{Y_b},$$

where Y = average simulated yield using one of the six cases specified in Table II and Y_b = baseline yield (36-year average yield using the observed 1960–1995 climate record).

For each crop, we used linear modeling techniques to test the hypothesis that mean yields from all scenarios and cases were equal. If this hypothesis was rejected, we conducted pairwise comparisons to test whether selected pairs of cases produced significantly different mean yields. Since standard analysis of variance

Table II
Soybean simulation scenarios

Case	Climate change scenario	Ambient CO_2 available to crop (ppm)	Soybean		Sorghum	
			Planting date	Cultivar	Planting date	Cultivar
Baseline (B)	1960–1995 observed	330	150	Generic MG V	150	Cargill 837
CSIRO climate change only (C)	CSIRO	330	150	Generic MG V	150	Cargill 837
RCM climate change only (R)	RCM	330	150	Generic MG V	150	Cargill 837
CSIRO climate change and elevated CO_2 (C2)	CSIRO	540	150	Generic MG V	150	Cargill 837
RCM climate change and elevated CO_2 (R2)	RCM	540	150	Generic MG V	150	Cargill 837
CSIRO climate change and elevated CO_2 w/adaptation (C2a)	CSIRO	540	Optimal	Optimal	100	Shortened juvenile and grain-fill periods
RCM climate change and elevated CO_2 w/adaptation (R2a)	RCM	540	Optimal	Optimal	100	Shortened juvenile and grain-fill periods

(ANOVA) methods require independent sample data, and simulated yields are correlated in both space and time, we used a mixed models approach (Littel et al., 1996) to account for spatial and temporal autocorrelation. We conducted separate analyses to determine which climate variables contributed most to differences between soybean yields simulated with the CSIRO and RCM scenarios. This included eight CSIRO grids that comprise approximately 95% of the study area. For each CSIRO grid, yield differences were regressed on climate variable differences during the growing season. Finally, we examined yield changes with regard to temperature and moisture stress during important physiological stages.

Table III
State-level dryland soybean yield changes for each scenario

State	Baseline (B)		% Change from baseline soybean yield and 36-year coefficient of variation											
	Yield	CV	CSIRO 330 (C)		RegCM 330 (R)		CSIRO 540 (C2)		RegCM 540 (R2)		CSIRO 540 + a (C2a)		RegCM 540 + a (R2a)	
	(t ha^{-1})		% Δ	CV	% Δ	CV	% Δ	CV	% Δ	CV	% Δ	CV	% Δ	CV
AL	2.21	28.2	–53.4	41.2	–77.7	45.8	–30.3	41.3	–66.1	47.2	–15.6	36.9	–18.8	22.2
AR	1.79	36.8	–66.4	50.6	–61.1	49.3	–51.3	50.3	–42.6	49.7	+6.9[a]	24.4	+13.4	19.6
FL	2.55	13.7	–35.4	27.1	–55.3	35.1	–3.8	27.2	–33.3	34.9	+2.4	25.3	–8.5	22.8
GA	2.34	24.7	–50.5	38.8	–78.6	41.2	–27.0	38.7	–68.1	41.3	–16.4	37.6	–33.6	23.6
LA	2.40	18.8	–42.4	28.6	–52.2	37.1	–16.8	27.5	–29.3	35.6	–1.7[a]	24.5	–1.7	26.8
MS	2.47	21.2	–47.4	36.4	–71.6	49.6	–23.9	36.0	–57.8	49.8	–7.9	36.4	–15.9	30.3
NC	2.69	18.0	–51.0	32.9	–73.7	35.2	–28.6	32.7	–61.6	35.4	–15.2	26.2	–39.5	25.8
SC	2.71	22.8	–52.1	42.2	–81.6	45.5	–30.3	42.2	–73.5	47.6	–17.6	37.4	–43.6	28.7
TN	2.60	22.9	–43.4	42.3	–66.8	55.1	–17.7	40.2	–49.9	53.1	+1.0	16.7	–12.4	19.6
Domain	2.38	19.1	–49.4	31.2	–69.2	35.8	–26.1	30.8	–54.3	25.7	–7.6	19.7	–18.3	17.8

[a] Indicates that yields are NOT significantly different ($\alpha = 0.05$).

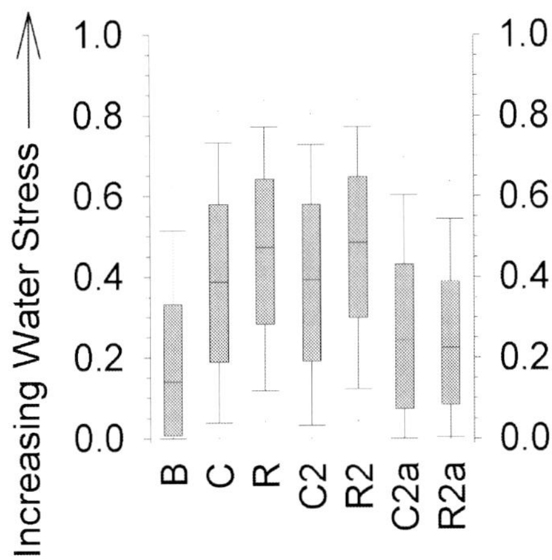

Figure 3. Soybean pod-fill water stress for each scenario and case. Growth rates are reduced linearly with increasing moisture stress from 0 (no stress) to 1 (complete stress).

There was essentially no difference in the yield results between climate change alone and climate changes with CO_2 fertilization. (3) Adaptation strategies (discussed below) generally reduced variability. For the CSIRO scenarios, variability modestly decreased in most states, but significantly decreased in Arkansas and Tennessee. For the RCM scenarios, variability significantly decreased in all states. While interannual yield variability in some states was higher for the adaptation cases than the baseline climate, it was the same or lower in many states.

The spatial pattern of yield decrease related closely to the greatest changes in maximum temperature and precipitation. These two variables combine to cause higher water stress during the pod-fill period in nearly every grid cell and year. Water stress during pod-fill significantly influences subsequent soybean yield (Haskett et al., 2000). CROPGRO-Soybean reduces photosynthesis and growth rates linearly as a function of computed water stress. Water stress during pod-fill is significantly higher for each of the climate change scenarios (Figure 3). It was associated with 44–68% of the simulated yield variance depending on scenario. Interestingly, CO_2 fertilization did not affect water stress. CO_2 enrichment increased leaf area index and, therefore, photosynthetic production. However, evapotranspiration remained essentially the same for the climate change only and elevated CO_2 scenarios because model adjustments for stomatal closure at higher CO_2 levels offset the influence of greater leaf area. When CO_2 was increased to 540 ppm, average plant biomass and yield increased by at least 20% in every grid cell. Consequently, water use efficiency, defined as either biomass or yield divided by evapotranspiration, increased for both climate change scenarios (Figure 4). This finding is consistent with Phillips et al. (1996).

Water stress during reproductive stages did not explain all low yields. In some cases, it was related to high temperatures during reproductive stages. CROPGRO-Soybean optimizes development rate (including processes related to photosynthetic

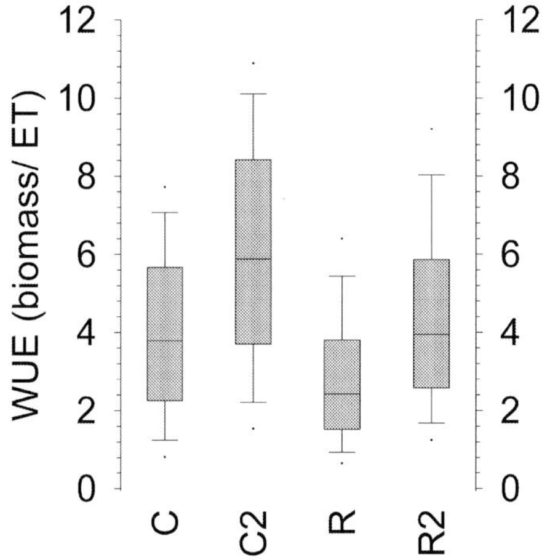

Figure 4. Soybean water use efficiency (WUE) as a function of CO_2 fertilization for CSIRO and RCM scenerios. WUE is measured as biomass or yield divided by evapotranspiration.

growth, node formation, seed growth, pod addition, and partition to pods) when hourly temperatures range between 22 and 34 °C, and it reduces these processes linearly at temperatures below and above these values (Boote et al., 1998). While it is impractical to reconstruct hourly temperature for each grid cell and growing season, several summary variables (e.g., average bloom or pod-fill temperature) can be used to identify periods of potential temperature stress that explain yield decrease. Average bloom-period maximum temperatures were nearly always lower than 34 °C in the baseline scenario, and usually exceeded 34 °C for the two climatic change scenarios (Figure 5). While average maximum temperature during bloom serves only as a proxy for modeled temperature stress, the relationship between this variable and final yield approximates the model function linking temperature to development rate (Boote et al., 1998). As average bloom-period maximum temperature increases (from the baseline, to CSIRO, to RCM scenarios) fewer and fewer cases fall within the optimum temperature range, and yield drops significantly. We also found that yield dropped when average pod-fill maximum temperature exceeded certain thresholds. In the RCM scenario, when yield fell below 1.00 t ha^{-1} with *low* water stress, it was often because pod-fill temperature exceeded 38 °C (Figure 6). These results echo the findings of Ferris et al. (1998) whose greenhouse experiments showed reductions in photosynthetic rate at high temperatures. Results from irrigated simulations further support the argument that high reproductive-stage temperatures suppress yield. In these simulations moisture stress was eliminated. Yield increased in nearly every grid, but percentage change varied as a function of maximum temperature. The greatest increases occur in the coolest regions, and the smallest increases occur in areas with the greatest maximum temperature changes. Regression analysis, designed to show differences between CSIRO and RCM soybean yields, reinforced many of the explanations

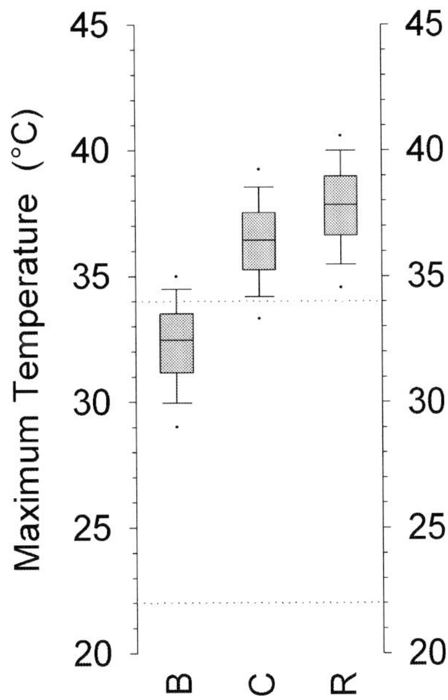

Figure 5. Soybean average bloom-period maximum temperature for each climate scenario.

above. July, August, and September precipitation, and August and September maximum and minimum temperature were the most consistent predictor variables accounting for yield differences between the CSIRO and RCM scenarios.

The very high temperatures during reproductive stages in the climate change scenarios also extended the average growing season length (defined as the period from germination to physiological maturity). This counter-intuitive result contrasts with some previous work (e.g., Alexandrov and Hoogenboom, 2000) showing that warmer temperatures hasten both vegetative and reproductive stages, thereby shortening the growing season. In our climate change scenarios, we find that growing season gets longer where temperature increases the most. For the CSIRO scenario, this occurs predominantly in the western part of the study area, while in the RCM scenario it is most pronounced in the eastern portions. Growing season was shortened only along the northern edge of the study region and at high elevations where baseline climate temperatures were below the optimal range for photosynthetic growth and, therefore, benefitted from the temperature increase imposed by the CSIRO and RCM scenarios. This point is illustrated by the relationship between growing season length and average maximum temperature during the bloom period (Figure 7). Below about 34 °C, growing season length generally decreases with increasing maximum temperatures during the bloom period. At temperatures higher than 34 °C, growing season length generally increases with increasing temperature. Average maximum temperature during most bloom periods in the baseline scenario are lower than 34 °C, but are higher than 34 °C in the CSIRO and RCM scenarios. These findings relate directly to the method by which CROPGRO-Soybean calcu-

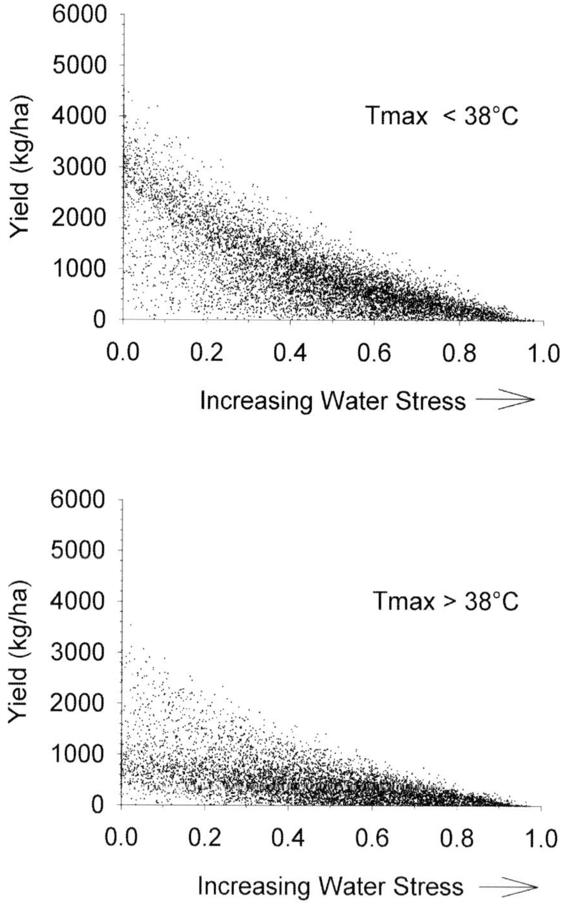

Figure 6. The relationship between pod-fill water stress and yield for the RCM, elevated CO_2 case and its dependence on temperature. Cases are separated on the basis of average pod-fill period maximum temperatures above or below 38 °C.

lates physiological development rates. Development in each stage is determined by sensitivity to temperature, photoperiod, and water and nitrogen stresses. Under optimal conditions for these variables, one physiological day is accumulated for each calendar day (Boote et al., 1989, 1998). Less than one physiological day was accumulated on every calendar day when temperature exceeded optimal conditions. Peart et al. (1989) and Curry et al. (1990) reported similar model behavior with high temperatures during reproductive stages. These findings, and the model structure itself, are further supported by both theoretical (Boote et al., 1998) and empirical (Ferris et al., 1999) studies examining the relationship between temperature and development rate.

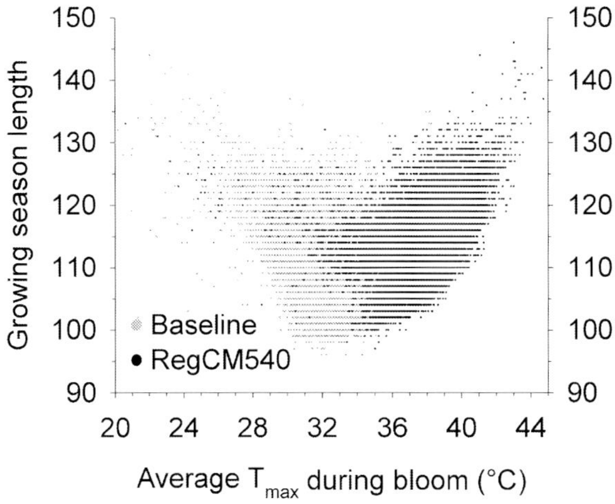

Figure 7. Growing season length vs. average bloom-period maximum temperature.

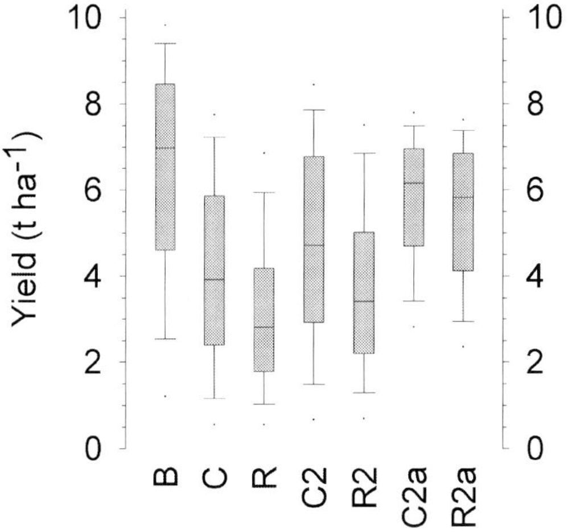

Figure 8. Dryland simulated sorghum yield for each climate scenario and case. Table II describes each case.

3.2. SORGHUM

Sorghum yield decreased significantly in response to the various climate change scenarios (Figure 8). In the climate change only cases, the CSIRO and RCM scenarios produced mean regional decreases of 36% and 51% respectively. Yield decreased from the baseline simulations in over 90% of all grid cells. As with soybean, interannual variability (measured by the coefficient of variation for state-averaged yield over the 36-year period) also increased in both climate change scenarios without adaptation (Table IV).

The spatial pattern of yield decreases closely matched that for soybean as both crops were influenced by the CSIRO and RCM patterns of maximum temperature

Table IV

State-level dryland sorghum yield changes for each scenario

State	Baseline (B)		% Change from baseline sorghum yield and 36-year coefficient of variation											
	Yield ($t\ ha^{-1}$)	CV	CSIRO 330 (C)		RegCM 330 (R)		CSIRO 540 (C2)		RegCM 540 (R2)		CSIRO 540 + a (C2a)		RegCM 540 + a (R2a)	
			% Δ	CV	% Δ	CV	% Δ	CV	% Δ	CV	% Δ	CV	% Δ	CV
AL	6.36	22.8	–40.2	27.2	–61.4	29.2	–30.5	26.0	–54.1	28.2	–6.5	15.9	–6.4	16.2
AR	5.33	29.7	–55.1	33.6	–45.8	34.0	–45.0	31.9	–34.3	32.1	+1.1[a]	21.2	+5.1[a]	18.7
FL	7.19	11.1	–24.8	16.3	–37.4	20.0	–17.0	14.2	–29.7	18.7	–16.4	15.0	–15.7	12.2
GA	6.62	17.8	–36.5	26.3	–60.9	25.0	–26.6	24.5	–53.3	23.5	–15.0	22.5	–25.4	26.2
LA	6.53	15.9	–35.0	18.1	–35.8	20.0	–27.3	16.4	–27.0	18.0	–18.4	16.5	–11.8	13.6
MS	6.94	19.0	–38.8	25.9	–55.5	31.6	–29.5	23.5	–46.8	30.1	–18.0	15.1	–12.0	13.7
NC	6.26	14.7	–28.0	22.7	–51.7	23.3	–15.9[a]	21.0	–42.5	22.2	–3.9[a]	18.8	–28.5	28.2
SC	7.52	14.1	–36.9	25.8	–63.3	25.4	–28.1	23.6	–56.9	24.7	–23.3	24.3	–43.0	31.7
TN	5.94	20.0	–22.5	28.6	–41.5	32.0	–6.9[a]	24.9	–27.8	30.3	+9.3	13.6	–5.0[a]	19.0
Domain	6.41	13.8	–35.9	20.5	–50.9	21.6	–25.7	18.9	–41.9	20.4	–9.9	13.7	–15.2	14.5

[a] Indicates that yields are NOT significantly different ($\alpha = 0.05$).

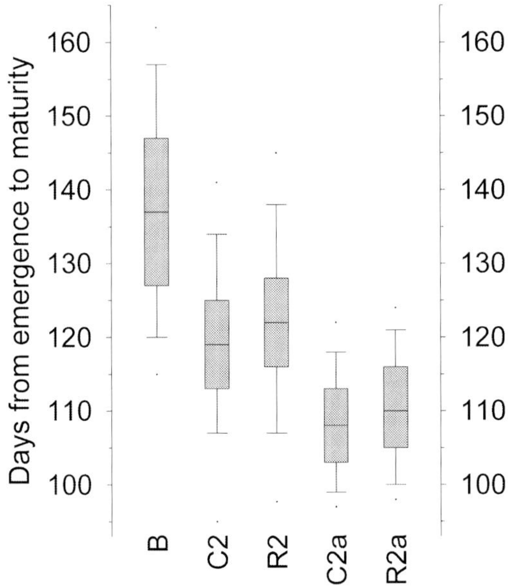

Figure 9. Sorghum growing season length for each climate scenario and case.

increases and precipitation decreases. The areas that fared best were the Gulf Coast, the north-central portion of the study region, and several Appalachian grids (where yield actually increased). The Gulf Coast had more modest temperature and precipitation changes than the rest of the Southeast. In the north-central and Appalachian grids, temperature increases had a less deleterious affect because baseline temperatures were lower than in other parts of the region. Regression analysis showed that July temperature and July and August precipitation were the most important predictor variables distinguishing the yield differences between CSIRO and RCM scenarios. The higher temperatures found in both the CSIRO and RCM scenarios shortened the grain-fill period and growing season by approximately 7 and 15 days, respectively (Figure 9). This finding was similar to that of Singh et al. (1998). When higher maximum temperatures were combined with lower precipitation, water stress also increased during the grain-fill period. Grain-fill water stress under the climate change cases was higher than the baseline case in nearly every year and grid cell (Figure 10).

Yield decreases were mitigated somewhat in the elevated-CO_2 cases, but not as dramatically as with soybean. This was expected, given that sorghum is a C_4 plant, and model adjustments relating ambient CO_2 concentration to photosynthetic rate differ between CERES-Sorghum and CROPGRO-Soybean accordingly. Nonetheless, CO_2 fertilization increased sorghum yield in nearly every year and grid cell (Figure 11). While an accelerated photosynthetic rate increased leaf area, stomatal resistance decreased in the simulation, limiting water losses. Consequently, water-use efficiency increased and grain-fill water stress was reduced in the elevated-CO_2

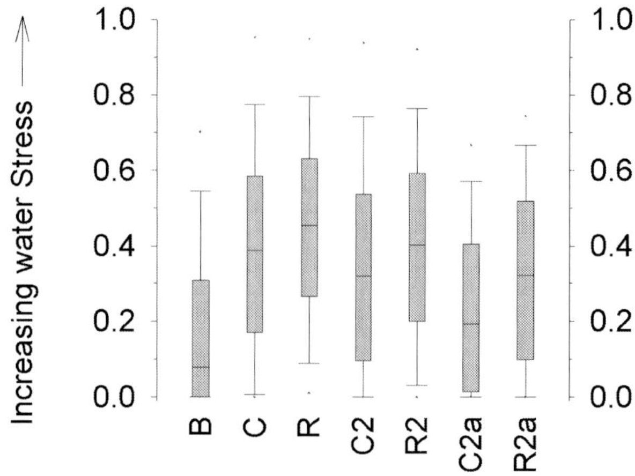

Figure 10. Sorghum grain-fill water stress for each scenario and case. Growth rates are reduced linearly with increasing moisture stress from 0 (no stress) to 1 (complete stress).

cases. This finding agrees with field studies of other C_4 species (Samarakoon and Gifford, 1996).

3.3. YIELD RESPONSE IN OTHER US REGIONS

We also simulated soybean and sorghum yield outside the Southeast using output from the same CSIRO simulations and RCM simulations for the western U.S. (Giorgi et al., 1998) and Great Lakes region (Bates et al., 1995). Output from these crop simulations was used in the agricultural sector model (ASM; see Adams et al., 2003) to measure net U.S. productivity in response to the two climate change scenarios. Soybean yield increased with the CSIRO and RCM climate changes projected in the Great Plains as precipitation increased during the spring and summer months. In the southern Great Lakes region, soybean yield generally increased, responding to the CSIRO increases in June and July precipitation, but decreased with the precipitation decreases of the RCM scenario. Sorghum yield increased in Nebraska and the Texas High Plains with the CSIRO scenario, but decreased in Kansas, Nebraska, Oklahoma, and the eastern half of Texas. These responses correspond closely with precipitation changes in June, July, and August. In the RCM scenario, precipitation increases caused modest sorghum yield increases across most of the Southern Plains. The only exception was in the Texas High Plains where RCM precipitation decreased, reducing yield. Given the importance of these soybean- and sorghum-growing regions, yield increases outside the Southeast influenced the results of the agricultural sector model.

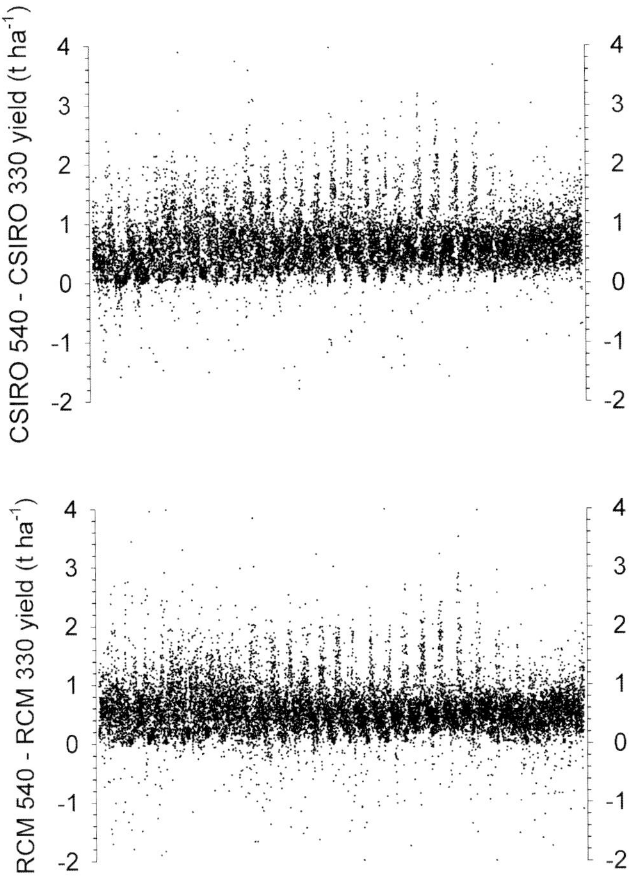

Figure 11. Sorghum yield differences between climate change only and climate change with elevated CO_2 for CSIRO and RCM climate change scenarios.

3.4. ADAPTATION CASES

3.4.1. *Soybean*

Adjusting planting date and cultivar significantly reduced the impact of climate change on simulated soybean yield. In the elevated-CSIRO case, average soybean yield increased from 1.76 t ha^{-1} to 2.20 t ha^{-1} when planting date and cultivar were altered in order to maximize state-level yield. Such adjustment produced higher yield than the baseline case in about half of the simulated years and grid cells, but because of extremely low yields in specific years and grid cells, domain-average yield was still 7.6% lower than the baseline case. Optimizing planting date and cultivar also improved yield response to the elevated RCM case. Simulated yield increased from 1.09 t ha^{-1} using the baseline planting date (calendar day 150) and cultivar (maturity group V), to 1.94 t ha^{-1} using dates and cultivars that optimized state-level yield. Given the more extreme temperature and precipitation changes of the RCM scenario, adaptation yield was higher than the baseline simulations in fewer than a third of all years and grid cells; regional average yield was still 18.3% lower than in the baseline simulations. Interannual variability also decreased in the adaptation levels such that it was approximately the same as the baseline scenario (Table III).

While optimal adjustments in planting date and maturity group varied spatially as a function of the intensity of each climate change scenario and soil type, two general strategies mitigated yield decreases due to climate change: planting lower maturity groups (I or IV) earlier in the season (mid-April) or planting higher maturity groups (V or VII) later in the season (mid-June). For the CSIRO scenarios, the former strategy improved yield in nearly half of all grid cells. It was particularly effective where the CSIRO-prescribed temperature and precipitation changes were greatest (Figure 12). In other regions, the highest yields resulted from the latter strategy – planting maturity group V or VII in mid-June. For the RCM scenarios, early planting with a low maturity group also reduced yield decreases. Seventy-five percent of all grid cells had the highest 36-year average yields using April 10 (calendar day 100) and maturity group (I or IV). Maturity group I worked best in the northern half of the domain, while maturity group IV dominated the southern half (Figure 12). Boote (1981) demonstrated that lower maturity groups could produce acceptable yield in the Southeast when planted early in the season. Yield in a number of grids benefitted from a shift to maturity group VII with a June planting date. This strategy worked best along the Coastal Plain of the Carolinas.

Two grid cells help to illustrate how planting and cultivar adaptations mitigate yield decreases associated with CSIRO and RCM climate change scenarios. One is located in Arkansas along the Mississippi River. The grid falls within a CSIRO cell where temperature increased between 3 and 6 °C from June through September, and precipitation decreased by 25–50% in July and August. The 36-year average CSIRO yield decreased by nearly 60% without adaptation. Planting maturity group IV earlier in the season (calendar day 100) provided one of the best adaptation strategies for this grid, as reduced pod-fill water stress and increased seed number and weight improved yield considerably. Despite the adjustments, the 36-year average was still 11% lower than the baseline yield. A second grid, located on the Coastal Plain of South Carolina, provides a less optimistic example. The projected RCM growing-season maximum temperature increase in this cell was over 8 °C; projected precipitation decreases ranged from 25 to 50%. Without adjustment, yield dropped by 70%. While no strategy could avoid the impacts of such changes, it is instructive to examine how early planting with a lower maturity group, or delayed planting with a higher maturity group, affected the timing of phenological stages relative to temperature. Figure 13 shows that the prolonged period of extreme temperatures would affect important growth stages for a wide range of adjustments. Planting quick-maturing varieties early, or slower-maturing varieties late in the growing season reduced heat and moisture stress, but yield was still 50% lower than in the baseline scenario. In this case, the RCM climate change

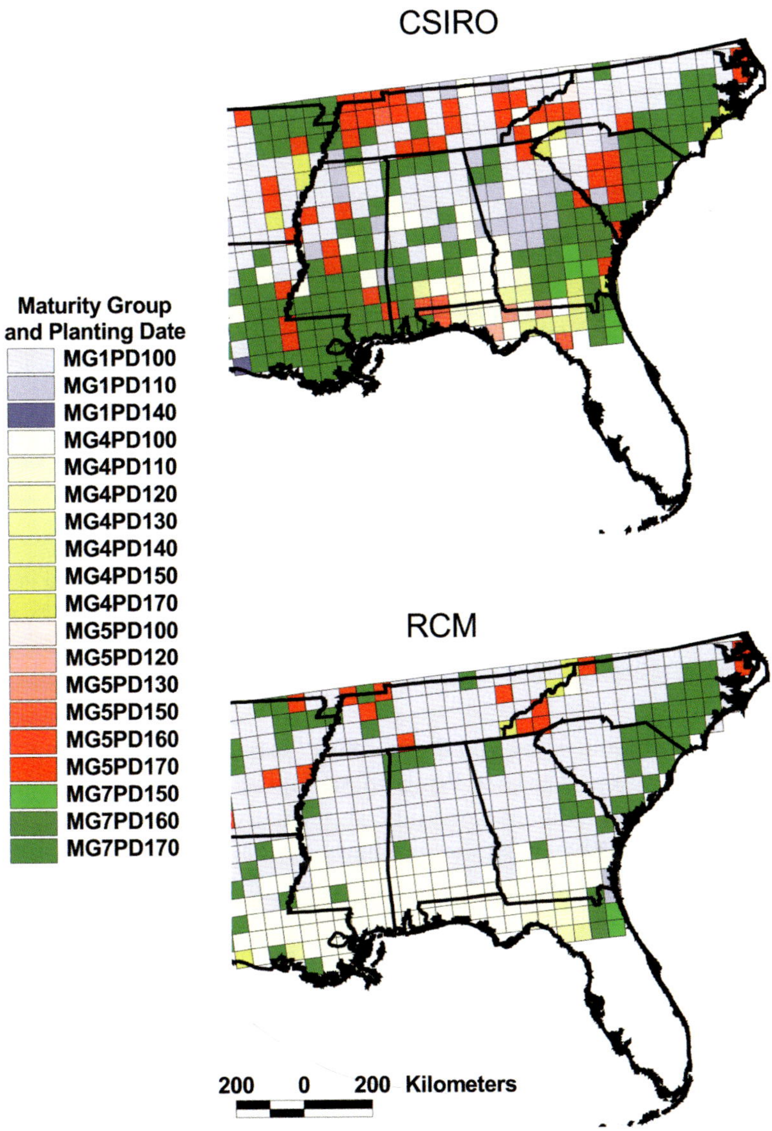

Figure 12. Optimal maturity group and planting date for the CSIRO and RCM climate change scenarios.

scenario presented extremes that the adaptation strategies considered here did not overcome.

3.4.2. *Sorghum*

Shortening the juvenile and grain-filling stages and advancing planting date reduced the impact of climate change on sorghum yield. In the elevated-CSIRO case, average yield increased from 4.76 t ha^{-1} to 5.78 t ha^{-1} when the thermal time of the

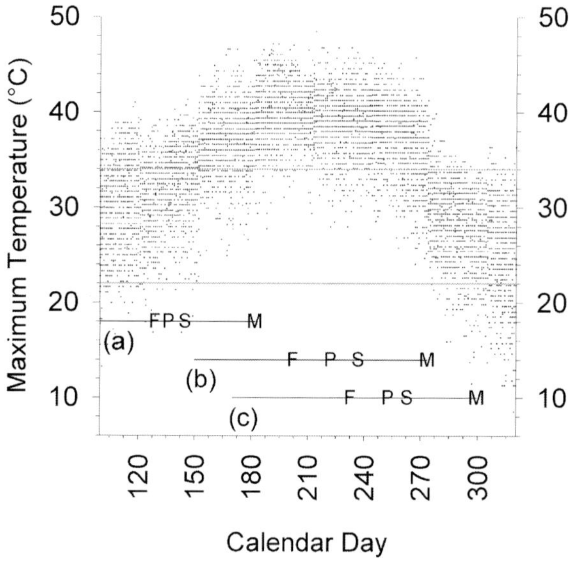

Figure 13. Maximum temperatures in the 36-year RCM time series for a 50-km grid in the South Carolina Coastal Plain. Dates of the first flower (F), pod (P), seed (S), and maturity (M) are shown for three adaptation strategies: (a) maturity group I, April 10 sowing; (b) maturity group V, May 20 sowing; and (c) maturity group VII, June 19 sowing.

juvenile stage was reduced by 100 degree days (base 8 °C) and planting date was shifted from May 30 to April 10. While these adjustments produced higher yield than the baseline case in about half of the simulated years and grid cells, average yield was still 9.9% lower than the baseline case. Optimizing planting date and cultivar also improved yield response to the elevated RCM case. Simulated yield increased from 3.73 t ha^{-1} to 5.44 t ha^{-1} with adjustments, but the Southeast average sorghum yield remained 15.2% lower than the baseline simulations. Equally important, our adaptation strategies decreased the interannual variability of yield such that it was comparable in magnitude to the baseline case. As with soybean, the adjustments to sorghum planting date and threshold thermal periods improved yield (Figure 8) by shifting the growing season to avoid water stress during grain fill (Figure 10). Unlike with soybean, however, the grain-fill period was shortened, reducing yield and countering some of the benefits of the adaptations.

4. Summary

This study examined how the spatial scale of climate change scenarios influences estimates of climate change impacts. Both the general circulation model and regional climate model used to create climate change scenarios for the southeastern United States projected temperature and precipitation changes that were associated with significant decreases in simulated soybean and sorghum yield. While consideration for direct CO_2 fertilization and for adjustments in planting date and cultivar mitigated yield decreases, regional averages remained lower than those simulated under baseline (1960–1995) conditions. Yield response to the projected climate changes was explained with respect to important physiological processes captured by the CROPGRO-Soybean and CERES-Sorghum models. Both crops were clearly influenced by water stress during important phenological stages (e.g., pod fill and grain fill). Successful adaptation strategies minimized water stress during these stages. Some of our results differ from other impact studies in the Southeast: for example, high maximum temperature slowed photosynthetic production and increased growing season length of soybean. This finding illustrates how certain thresholds, built into CROPGRO-Soybean, influence final yield in a manner consistent with other theoretical and empirical studies.

The spatial scale of climate change scenarios matters considerably. The magnitude and spatial patterns of yield response differed significantly between climate change scenarios. Again, such differences can be explained in the context of plant response to specific climate changes. The magnitude of temperature and precipitation projections varied between the CSIRO and RCM scenarios, producing different patterns of moisture stress in the two scenarios. Moreover, differences in the timing of temperature and precipitation change influenced yield response since soybean and sorghum showed sensitivity to water or temperature stress during particular growth stages. The influence of scale extends to adaptation strategies. In some cases, yield was maximized with dramatically different planting date and cultivar choices depending on climate change scenario. Our findings suggest that magnitude of uncertainty associated with the spatial scale of climate change projections warrants full consideration in climate impact studies.

Acknowledgements

We thank Tim Haas for providing statistical advice, Cynthia Hays for constructing the spatial soils data base, and Jennifer Rainman for assisting with crop simulations and graphics. We also thank Karen Beidel and two anonymous reviewers for their useful comments on earlier drafts. This research was funded in part by the United States Environmental Protection Agency through R824997-01-0. However, it has not been subjected to the Agency's required peer and policy review and therefore does not necessarily reflect the views of the Agency; no official endorsement should be inferred. This research was also funded by the National Aeronautics and Space Administration, MTPE (Grant No. OA99073, W-19, 080), and the United States Department of Agriculture, NRICGP (Grant No. 98-35106-6837).

References

Adams, R. M., McCarl, B. A., and Mearns, L. O.: 2003, 'The Economic Effects of Spatial Scale of Climate Scenarios: An Example from U.S. Agriculture', *Clim. Change* **60**, 131–148.

Alexandrov, V. A. and Hoogenboom, G.: 2000, 'Vulnerability and Adaptation Assessments of Agricultural Crops under Climate Change in the Southeastern U.S.A.', *Theor. Appl. Climatol.* **67**, 45–63.

Bates, G. T., Hostetler, S. W., and Giorgi, F.: 1995, '2-Year Simulation of the Great-Lakes Region with a Coupled Modeling System', *Mon. Wea. Rev.* **123** (5), 1505–1522.

Baumer, O., Kenyon, P., and Bettis, J.: 1994, 'Prediction of Soil Properties', in *Map Unit Use Files (MUUF), User's Manual Version 2.14. Natural Resources Conservation Service*, Wetland Science Institute, Laurel, MD.

Birch, C. J., Carberry, P. S., Muchow, R. C., McCown, R. L., and Hargreaves, J. N. G.: 1990, 'Development and Evaluation of a Sorghum Model Based on CERES-Maize in a Semi-Arid Tropical Environment', *Field Crops Res.* **24**, 87–104.

Boote, K. J.: 1981, 'Response of Soybeans in Different Maturity Groups to March Plantings in Southern U.S.A.', *Agronomy J.* **73**, 854–859.

Boote, K. J., Jones, J. W., and Hoogenboom, G.: 1989, 'Simulating Growth and Yield Response of Soybean to Temperature and Photoperiod', in *Proceedings of the World Soybean Research Conference IV*, 5–9 March 1989, Buenos Aires, Argentina, pp. 273–278.

Boote, K. J., Jones, J. W., Hoogenboom, G., and Pickering, N. B.: 1998, 'The CROPGRO Model for Grain Legumes', in Tsuji, G. Y., Hoogenboom, G., and Thornton, P. K. (eds.), *Understanding Options for Agricultural Production*, Kluwer Academic Publishers, Dordrecht, pp. 99–128.

Boote, K. J., Jones, J. W., Hoogenboom, G., and Wilkerson, G.: 1997, 'Evaluation of the CROPGRO-Soybean Model over a Wide Range of Experiments', in Teng, P. S., Kropff, M. J., Berge, H. F. M., Dent, J. B., Lansigna, F. P., and van Laar, H. H. (eds.), *Applications of Systems Approaches at the Field Level*, Kluwer Academic Publishers, Dordrecht, pp. 113–133.

Boote, K. J., Jones, J. W., and Pickering, N. B.: 1996, 'Potential Uses and Limitations of Crop Models', *Agronomy J.* **88**, 704–716.

Boote, K. J. and Pickering, N. B.: 1994, 'Modeling Photosynthesis of Row Crop Canopies', *Hort Science* **29**, 1423–1434.

Brisson, N., Bona, S., and Bouniols, A.: 1989, 'SOYGRO, un Modèle de Simulation de la Culture du Soja; Adaptation à des Variétés dans le Sud de L' Europe et Validation', *Agronomie* **9**, 27–37.

Brown, R. A., Rosenberg, N. J., Hays, C. J., Easterline, W. E., and Mearns, L. O.: 2000, 'Potential Production and Environmental Effects of Switchgrass and Traditional Crops under Current and Greenhouse-altered Climate in the Central United States: A Simulation Study', *Agriculture, Ecosystems and Environment* **78**, 31–47.

Carbone, G.J., Mearns, L.O., Mavromatis, T., Sadler, E.J., and Stooksbury, D.: 2003, 'Evaluating CROPGRO-Soybean Performance for use in Climate Impact Studies', *Agronomy J.* **95**, 537–544.

Castrignano, A., Di Bari, V., and Stelluti, M.: 1997, 'Evapotranspiration Predictions of CERES-Sorghum Model in Southern Italy', *European J. Agronomy* **6**, 265–274.

Colson, J., Bouniols, A., and Jones, J. W.: 1995, 'Soybean Reproductive Development: Adapting a Model for European Cultivars', *Agronomy J.* **87**, 1129–1139.

Curry, R. B., Jones, J. W., Boote, K. J., Peart, R. M., Allen, L. H., and Pickering, N. B.: 1995, 'Response of Soybean to Predicted Climate Change in the U.S.A.', in *Climate Change and Agriculture: Analysis of Potential International Impacts*, ASA Special Publication No. 59.

Curry, R. B., Peart, R. M., Jones, J. W., Boote, K. J., and Allen, L. H.: 1990, 'Response of Crop Yield to Predicted Changes in Climate and Atmospheric CO_2 Using Simulation', *Transactions of the American Society of Agricultural Engineers* **33** (4), 1383–1390.

Easterling, W. E., Mearns, L. O., Hays, C. J., and Marx, D.: 2001, 'Comparison of Agricultural Impacts of Climate Change Calculated from High and Low Resolution Climate Change Scenarios: Part II, Accounting for Adaptation and CO_2 Direct Effects', *Clim. Change* **51** (2), 173–197.

Ferris, R., Wheeler, T. R., Ellis, R. H., and Hadley, P.: 1999, 'Seed Yield after Environmental Stress in Soybean Grown under Elevated CO_2', *Crop Science* **39**, 710–718.

Ferris, R., Wheeler, T. R., Hadley, P., and Ellis, R. H.: 1998, 'Recovery of Photosynthesis after Environmental Stress in Soybean Grown under Elevated CO_2', *Crop Science* **38**, 948–955.

Giorgi, F., Mearns, L., Shields, C., and McDaniel, L.: 1998, 'Regional Nested Simulations of Present Day and $2 \times CO_2$ Climate over the Central Great Plains of the United States', *Clim. Change* **40**, 457–493.

Guerena, A., Ruiz-Ramos, M., Diaz-Ambrona, C. H., Conde, J. R., and Minguez, M. I.: 2001, 'Assessment of Climate Change and Agriculture in Spain Using Climate Models', *Agronomy J.* **93**, 237–249.

Hammer, G. L. and Muchow, R. C.: 1994, 'Assessing Climatic Risk to Sorghum Production in Water-Limited Subtropical Environments I. Development and Testing of a Simulation Model', *Field Crops Res.* **36**, 221–234.

Haskett, J. D, Pachepsky, Y. A., and Acock, B.: 2000, 'Effect of Climate and Atmospheric Change on Soybean Water Stress: A Study of Iowa', *Ecol. Modelling* **135**, 265–277.

Hoogenboom, G., Jones, J. W., Wilkens, P. W., Batchelor, W. D., Bowen, W. T., Hunt, L. A., Pickering, N. B., Singh, U., Godwin, D. C., Baer, B., Boote, K. J., Ritchie, J. T., and White, J. W.: 1994, 'Crop Models', in Tsuji, G. Y., Uehara, G., and Balas, S. (eds.), *DSSAT v3.*, University of Hawaii, Honolulu, Hawaii, pp. 95–244.

Houghton, J. T., Ding, Y., Griggs, J. J., Noguer, M., van der Linden, P. J., Dai, X., Maskell, K., and Johnson, C. A. eds.: 2001, *IPCC, 2001: Climate Change 2001: The Scientific Basis. Contribution of Working Group I to the Third Assessment Report of the Intergovernmental Panel on Climate Change*, Cambridge University Press, New York, 881 pp.

Kiniry, J. R., Williams, J. R., Vanderlip, R. L., Atwood, J. D., Reicosky, D. C., Mulliken, J., Cox, W. J., Mascagni, H. J., Hollinger, S. E., and Wiebold, W. J.: 1997, 'Evaluation of Two Maize Models for Nine U.S. Locations', *Agronomy J.* **89**, 421–426.

Littel, R. C., Milliken, G. A., Stroup, W. W., and Wolfinger, R. D.: 1996, *SAS System for Mixed Models*, SAS Institute Inc., Cary, NC, 633 pp.

Mearns, L. O., Easterling, W., Hays, C., and Marx, D.: 2001, 'Comparison of Agricultural Impacts of Climate Change Calculated from High and Low Resolution Climate Change Scenarios: Part I. The Uncertainty due to Spatial Scale', *Clim. Change* **51** (2), 131–172.

Mearns, L. O., Giorgi, F., McDaniels, L., and Shields, C.: 2003a, 'Climate Scenarios for the Southeast U.S. based on GCM and Regional Model Simulations', *Clim. Change* **60**, 7–35.

Mearns, L.O., Carbone, G., Doherty, R. M., Tsvetsinskaya, E. A., McCarl, B. A., Adams, R. M., and McDaniel, L.: 2003b, 'The Uncertainty Due to Spatial Scale of Climate Scenarios in Integrated Assessments: An Example from U.S. Agriculture', *Integrated Assessment*, accepted.

Peart, R. M., Jones, J. W., Curry, R. B., Boote, K. J., and Allen, L. H.: 1989, 'Impact of Climate Change on Crop Yield in the Southeastern U.S.A.: A Simulation Study', in Smith, J. B. and Tirpak, D. A. (eds.), *The Potential Effects of Global Climate Change on the United States*, USEPA-230-05-89-053. Appendix C. Agriculture, Vol. 1. USEPA, Washington, D.C., pp. 1–54.

Phillips, D. L., Lee, J. L., and Dodson, R. F.: 1996, 'Sensitivity of the U.S. Corn Belt to Climate Change and Elevated CO_2: I. Corn and Soybean Yields', *Agricultural Systems* **52**, 481-502.

Pickering, N. B., Jones, J. W., and Boote, K. J.: 1995, 'Adapting SOYGRO v5.42 for Prediction under Climate Change Conditions', in Rosenzweig, C. E., Allen L. H. Jr., Harper, L. A., Hollinger, S. E., and Jones, J. W. (eds.), *Proceedings of a Symposium on Climate Change and International Impacts*, American Society of Agronomy Special Publication Number 59. American Society of Agronomy, Madison, WI, pp. 77–98.

Priestley, C. H. B. and Taylor, R. J.: 1972, 'On the Assessment of Surface Heat Flux and Evaporation Using Large-scale Parameters', *Mon. Wea. Rev.* **100**, 81–92.

Reybold, W. U. and TeSelle, G. W.: 1989, 'Soil Geographic Data Bases', *J. Soil Water Conservation* **44**, 28–29.

Richardson, C. W. and Nicks, A. D.: 1990, 'Weather Generator Description', in *EPIC – Erosion/Productivity Impact Calculator. 1. Model documentation*, Agricultural Research Service, United States Department of Agriculture, Washington, D.C., pp. 93–104.

Robertson, M. J., Fukai, S., Hammer, G. L., and Ludlow, M. M.: 1993, 'Modelling Root Growth of Grain Sorghum using the CERES Approach', *Field Crop Res.* **33**, 113–130.

Rosenzweig, C. and Iglesias, A.: 1998, 'The Use of Crop Models for International Climate Change Impact Assessment', in Tsuji, G. Y., Hoogenboom, G., and Thornton, P. K. (eds.), *Understanding Options for Agricultural Production*, Kluwer Academic Publishers, Dordrecht, pp. 267–292.

Samarakoon, A. B. and Gifford, R. M.: 1996, 'Elevated CO_2 Effects on Water Use and Growth of Maize in Wet and Drying Soil', *Aust. J. Plant. Physiol.* **23** (1), 53–62.

Semenov, M. A. and Barrow, E. M.: 1997, 'Use of a Stochastic Weather Generator in the Development of Climate Change Scenarios', *Clim. Change* **35** (4), 397–414.

Singh, B., El Maayar, M. Andre, P., Bryant, C. R., and Thouez, J.-P.: 1998, 'Impacts of a GHG-Induced Climate Change on Crop Yields: Effects of Acceleration in Maturation, Moisture Stress and Optimal Temperature', *Clim. Change* **38**, 51–86.

Tsuji, G. Y., Jones, J. W., and Balas, S. (eds.), 1994, *DSSAT Version 3. A Decision Support System for Agrotechnology Transfer*, University of Hawaii, Honolulu.

Tubiello, F. N., Donatelli, M., Rosenzweig, C., and Stockle, C. O.: 2000, 'Effects of Climate Change and Elevated CO_2 on Cropping Systems: Model Predictions at Two Italian Locations', *European J. Agronomy* **13**, 179–189.

United States Department of Agriculture: 1997, Usual Planting and Harvesting Dates for U.S. Field Crops. Agricultural Handbook Number 628. http://usda.mannlib.cornell.edu/reports/nassr/field/planting/uph97.pdf

Wassenaar, T., Lagacherie, P., Legros, J.-P., and Rounsevell, M. D. A.: 1999, 'Modelling Wheat Yield Responses to Soil and Climate Variability at the Regional Scale', *Clim. Res.* **11**, 209–220.

Watterson, I. G.: 1998, 'An Analysis of the Global Water Cycle of Present and Doubled CO_2 Climates Simulated by the CSIRO General Circulation Model', *JGR-Atmospheres* **103** (D18), 23113–23129.

Watterson, I. G., Dix, M. R., and Colman, R. A.: 1999, 'A Comparison of Present and Doubled CO_2 Climates and Feedbacks Simulated by Three General Circulation Models', *JGR-Atmospheres* **104** (D2), 1943–1956.

Wolf, J.: 2002a, 'Comparison of Two Soya Bean Simulation Models under Climate Change. I. Model Calibrations and Sensitivity Analyses', *Clim. Res.* **20**, 55–70.

Wolf, J.: 2002b, 'Comparison of Two Soya Bean Simulation Models under Climate Change. II. Applications of Climate Change Scenarios', *Clim. Res.* **20**, 71–81.

Xie, Y., Kiniry, J. R., Nedbalek, V., and Rosenthal, W. D.: 2001, 'Maize and Sorghum Simulations with CERES-Maize, SORKAM, and ALMANAC under Water-Limiting Conditions', *Agronomy J.* **93**, 1148–1155.

(Received 8 May 2002; in revised form 31 March 2003)

SPATIAL SCALE EFFECTS OF CLIMATE SCENARIOS ON SIMULATED COTTON PRODUCTION IN THE SOUTHEASTERN U.S.A.

RUTH M. DOHERTY [1], LINDA O. MEARNS [1], K. RAJA REDDY [2], MARY W. DOWNTON [1] and LARRY McDANIEL [1]

[1] *Environmental and Societal Impacts Group, National Center for Atmospheric Research, P.O. Box 3000, Boulder, CO 80307, U.S.A.*
E-mail: rdoherty@ucar.edu

[2] *Department of Plant and Soil Sciences, Mississippi State University, MS, U.S.A.*

Abstract. We examine the effect of climate scenarios generated using results from climate models of different spatial resolution on yields simulated by the deterministic cotton model GOSSYM for the southeastern U.S.A. Two related climate change scenarios were used: a coarse-scale scenario produced from results of a general circulation model (GCM) which also provided the boundary conditions to a regional climate model (RCM), from which a fine-scale scenario was constructed. Cotton model simulations were performed for three cases: climate change alone; climate change and elevated CO_2; climate change, elevated CO_2 and adaptations to climate change. In general, significant differences in state-average projected yield changes between the coarse and fine-scale scenarios are found for these three cases. In the first two cases, different directions of change are found in some sub-regions. With adaptation, yields substantially increase for both climate scenarios, but more so for the coarse-scale scenario (30% domain-average increase). Under irrigation, yield change differences between the two climate scenarios are small in all three cases, and yields are higher under irrigation (~35% domain-average increase with adaptation case) compared to dryland conditions. For the climate change alone case, differences in summer water-stress levels explain the contrasts in dryland yield patterns between the coarse and fine-scale climate scenarios.

1. Introduction

Cotton production is very important economically for the United States, accounting for 25–30% of global trade in raw cotton (http://www.ers.usda.gov/briefing/cotton/). Cotton acreage in the United States rose slightly during the 1990s from the previous two decades, and cotton production has similarly increased during that time, alongside advances in technology (seed varieties, fertilizers, pesticides, and machinery), production practices (reduced tillage, irrigation, crop rotations, pest management systems) as well as increases in atmospheric CO_2 concentrations which may have contributed to enhanced production. In the 1970s and 1980s, cotton acreage averaged about 12 million acres and by 2000 it had surpassed 15 million acres. Cotton is grown widely across the southern U.S.A., for both dryland and irrigated conditions in the southeastern states and Texas, and under irrigation in the southwestern states including southern California.

Climatic Change **60:** 99–129, 2003.

CO_2, and climate change and elevated CO_2 including changes in management practices to adapt to climate change. Section 2 describes methods including climate scenario construction and the cotton model used, experiments performed and subsequent analysis. The projected changes in yields, phenological events and boll development under the different climate scenarios and different cases are discussed in Section 3. Multiple regression analyses are used in Section 4, to discern the most important climate and environmental parameters of spatial variability in mean yields, both at the southeastern domain and sub-domain level. Summary and conclusions are presented in Section 5.

2. Methods

2.1. CLIMATE SCENARIO CONSTRUCTION

A present-day baseline climate dataset and two climate change scenario datasets were constructed for the southeastern U.S.A. (approximately 29°–36° N, –95°–75° W). A complete description of the baseline climate data set and the two climate scenarios is presented in a companion paper (Mearns et al., 2003b). All of these climate datasets have a resolution of 50 km on an equal-area grid. The present-day dataset of daily climate variables of maximum temperature, minimum temperature, precipitation and solar radiation, representing a 36-year period from 1960–1995, was constructed from the co-operative network of station data. Extensive quality control was performed to select appropriate stations (Mearns et al., 2003b), and each gridded value represents the nearest reliable station to the centre of the grid. The grid employed was the equal area lat./lon. grid on which the regional climate model simulations (see below) were performed. One climate change scenario was constructed from simulations with the CSIRO Mk2 GCM which has a resolution of (3.2° × 5.6°) and 8 land grid-boxes over the southeastern domain and the other was constructed from the RegCM2 RCM (50 km resolution; 414 grid boxes over the southeastern domain) for which the CSIRO model provided the initial and lateral boundary conditions for the RegCM model. The CSIRO model was chosen as the driving model of the RCM as its control run showed a more reasonable representation of the present-day climatology than results from other GCMs that were available at the time when these RCM simulations commenced (Giorgi et al., 1998). In this standard nesting procedure, it is assumed that the GCM provides the climate response to increased CO_2 for the large-scale circulation and the RCM provides the sub-GCM grid scale response to the forcing. The two scenarios are obviously associated through the boundary conditions from the GCM. More recent analysis of inter-GCM model simulations of future change in precipitation under the IS92a and SRES scenarios show there to be an inconsistent sign in the direction of change over this region (Giorgi et al., 2001). For example, contrasts in changes in precipitation are seen in the simulations of the atmosphere-ocean GCM

(AOGCMs) used in the recent U.S. National Assessment (Doherty and Mearns, 1999; Felzer and Heard, 1999). For the southeast region at the end of the 21st century the HadCM2 AOGCM projected increases in precipitation of about 25% in summer, whereas the CGCM1 AOGCM projected decreases of up to 25%. Differences between the CSIRO and RegCM simulations are mainly a function of scale and increased topographical resolution at the 50 km grid with differences in model physics potentially having some role to play. Scale contrasts in the scenarios are described extensively in Mearns et al. (2003b) and Giorgi et al. (1998), as well as being outlined below.

Under both climate scenarios over the southeastern region precipitation increases in the spring, but these increases are higher in the GCM (~30%) compared to the RCM (~20%). By July, August and September there are large decreases in precipitation (~20%) in the GCM across most of the southeast domain (e.g., August, Figures 1a,b). The southern parts of the domain show smaller decreases of 0–10% in July and August (Figures 1a,b). The decreases in precipitation in the GCM become more extreme (~30%) in the RCM in summer especially along the eastern coast. As discussed by Mearns et al. (2003b) there can be large differences in spatial patterns of precipitation at the sub-domain scale between the GCM and RCM. This is particularly the case during the warm season when local convective processes play a relatively important role. Maximum temperature increases in the GCM tend to remain in the range of 3–5 °C, and in spring and summer, the southernmost regions have the smallest increases of 3–4 °C. In the RCM, temperature increases are much more spatially and temporally variable, ranging from 1–7 °C across growing season months. In summer for the RCM, large increases (up to 7 °C) in maximum temperatures are found in the northeastern part of the domain extending into Georgia where maximum drying occurs (e.g., August, Figure 1d), whereas the increases in the GCM are smaller (Figure 1c). Patterns of change in solar radiation (not shown) tend to be similar but inverted to those for precipitation. The largest increases occur in summer (up to 30%) along the southern and eastern coast in the RCM (see Mearns et al., 2003b). Overall, the changes in climate in the RCM are more extreme than in the GCM. See Mearns et al. (2003b) for further details of the changes in climate in the RCM and GCM.

Climate change scenarios of 36-year length were constructed by applying mean monthly climate differences (for temperature) or ratios (for precipitation, solar radiation and wind-speed) between doubled and single CO_2 simulations performed with the two models, to the present-day dataset. For the case of the RegCM RCM, hereafter the fine-scale scenario, a unique monthly value was applied to each of the 414 50-km grid boxes, whilst for the CSIRO GCM, hereafter the coarse-scale scenario, as discussed above only 8 GCM land grid boxes covered the domain. In the construction of the climate scenarios, we applied mean monthly changes in climate to the present-day dataset and hence changes in year-to-year variability were not incorporated.

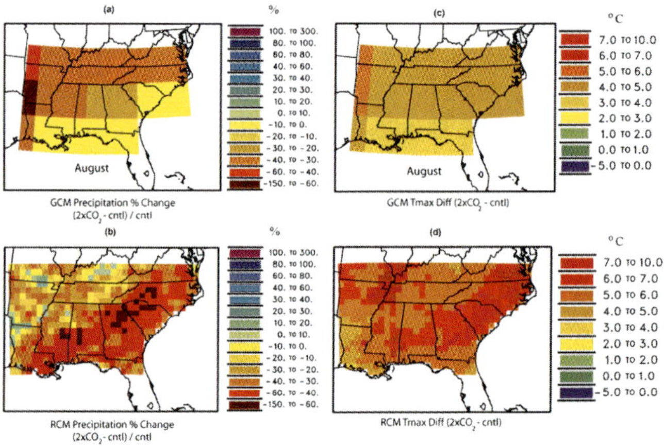

Figure 1.

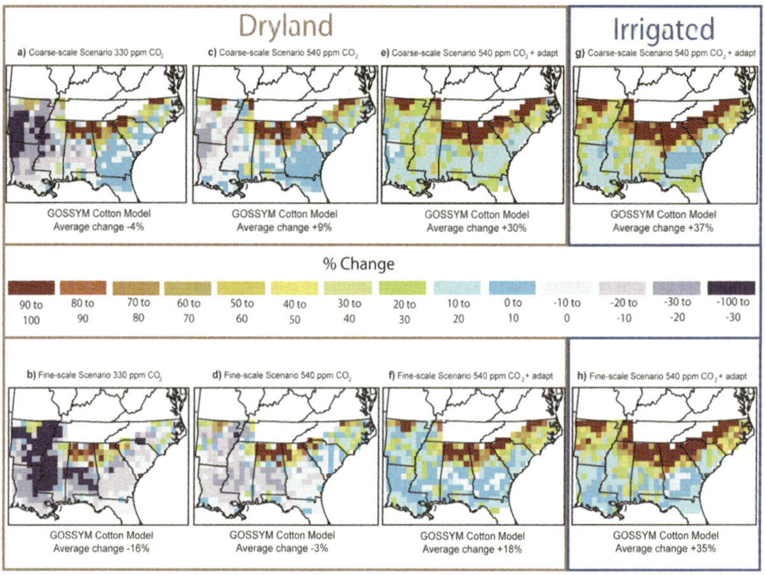

Figure 2.

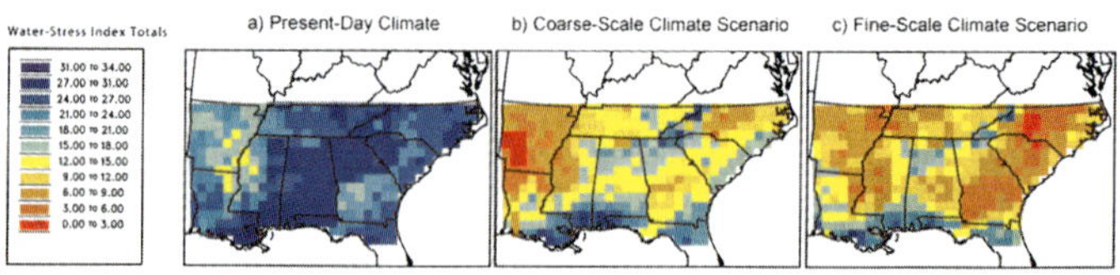

Figure 4.

Present-day station and climate change simulations from the CSIRO and RegCM climate models (Giorgi et al., 1998) were also used to produce climate scenarios for other cotton growing areas in Texas and the southwestern U.S.A. (New Mexico, Arizona) across to Southern California where cotton is grown mainly under irrigation. The same control and doubled CO_2 experiments of the CSIRO GCM were used, and the same version of the RegCM as for the southeast climate simulations. This was done to provide yield estimates for all regions in the U.S. where cotton is currently grown for use in the Agriculture Sector Model (ASM) economic model (see companion paper Adams et al., 2003).

The regions were selected according to county acreage maps of cotton production (http://www.usda.gov/nass/graphics/county00/ctuyld.htm, USDA, 2000). We briefly describe these southwestern scenarios below and describe yield results in Section 3.2 for completeness. However, this paper focusses primarily on results over the southeastern U.S.A. region.

For the GCM, in spring northern and eastern Texas and Oklahoma tend to experience small precipitation increases, but Arizona and New Mexico see precipitation decreases. By summer, precipitation decreases tend to be uniform across the southwest, except for southern California, where there is little change. Precipitation changes in the RCM are broadly similar to those of the GCM, but differ in local details. In spring, mainly precipitation increases are found in the RCM, particularly in the northern and eastern parts of Texas that are larger than those found in the GCM. In later summer months, areas of large decreases ($\sim$35%) are found over northern Texas, New Mexico and Arizona. The decreases in precipitation in summer in the GCM however are in general larger than those in the RCM. Temperature changes are fairly uniform in the GCM, around 4 °C (and slightly lower for southern California) for most seasons, except in summer (around 5 °C). Temperature contrasts are not large between the two models, except in summer when the increases are larger for the RCM.

Facing page

Figure 1. August percentage changes in precipitation (a, *b*) and differences in maximum temperature (c, *d*) simulated by the GCM and *RCM* respectively between the doubled CO_2 and control simulations.

Figure 2. Percentage changes in simulated yields between the coarse and *fine*-scale climate scenarios and present-day for the three cases for dryland conditions (coarse: a, c, e, and *fine: b, d, f*) and for the adaptation case – for irrigated conditions (g, h).

Figure 4. August water-stress index totals over the Southeastern U.S., for present-day (a) and the coarse (b) and fine-scale (c) scenarios for climate change alone case. Daily water-stress index values are accumulated over the month. A high index total indicates low water-stress and a low index total indicates high levels of water-stress.

2.2. THE COTTON GROWTH MODEL: GOSSYM

Cotton growth and yields are simulated by GOSSYM (Reddy et al., 1997a,b,c; Hodges et al., 1998), which is a detailed *mechanistic* crop model that accounts for carbon, nitrogen, and water processes in the plant and soil root-zone. It has a fully dynamic soil sub-model with multiple layers that vary with soil type, and an interactive rooting zone. GOSSYM operates on a daily time-step and simulates plant photosynthesis, respiration, and the accumulation of above and below-ground biomass. Biomass is partitioned to each plant organ in proportion to growth requirements, essentially the ratio of carbohydrate supply to demand (Hodges et al., 1998), and consequently plant organ growth is simulated through different phenological stages. GOSSYM has been developed and continuously updated with results most recently from elevated temperature and CO_2 experiments conducted in sunlit environment-controlled plant growth chambers (Reddy et al., 1997a,b, 2000, 2002 and references cited therein). GOSSYM simulates crop response to climate variables (temperature, rain/irrigation, solar radiation and wind) as well as environmental variation in soil and cultural practices. Model deficiencies include not having a detailed energy-balance model (a limitation of many crop models), and not accounting for relative humidity effects. However, evapotranspiration does use vapour pressure deficit as a function of minimum and average temperature.

The GOSSYM model has been extensively validated for present-day conditions across a wide range of environmental and cultural practices at numerous sites in the U.S. cottonbelt; Mississippi (Reddy et al., 1985), Texas (Staggenborg et al., 1996), Missouri (Albers et al., 1992) and several other southern states (Boone et al., 1993; Reddy et al., 1995; Reddy and Boone, 2002). However, GOSSYM does not work as well for irrigated cotton in arid environments (Fye et al., 1985).

2.3. ENVIRONMENTAL INPUT DATASETS AND CASES

Environmental inputs to the GOSSYM model included soil types (with associated hydrological and optical parameters), planting and harvest dates, variety types, and management practices. Representative soil texture types for the southeastern region were selected on the basis of the United Nations/FAO soil texture class dataset (http://www.fao.org/ag/AGL/agll/dsmw) at a 5 minute resolution and the 30 second STATSGO soils texture dataset (http://www.ftw.nrcs.usda.gov/stat_data.html). GOSSYM requires detailed soil inputs that describe the physical characteristics of the soil that are used by the model. These inputs include detailed parameters concerning plant available water and water movement through the soil and GOSSYM has a database of these for a number of soil types. These soil inputs are more detailed and numerous than those generally required by other crop models such as CERES (see Carbone et al., 2003; Tsvetsinkaya et al., 2003) and EPIC (Easterling et al., 2003), and so we were not able to use the soil databases developed for use in the other crop models in this southeastern study).

Thus, from the GOSSYM model soil database, three soil types were chosen to generally represent the appropriate soils types for the region at 50-km resolution:

- A (Norfolk) loamy sand (2 layers: depths – 20 and 201 cm, available water – 0.15 and 0.09 cc/cc) covers the coastal areas of North and South Carolina, Georgia, and Florida.
- A (Bosket3) sandy loam (3 layers: depths – 28, 61 and 201 cm, available water – 0.24, 0.20, 0.26 cc/cc) covers the interior of the domain (northern South Carolina, the northern half of Georgia, Alabama, the eastern half of Mississippi and northwestern Louisiana).
- A (Dundee1) silt loam (3 layers: depths – 15, 35 and 201 cm, available water – 0.15, 0.12, 0.11 cc/cc) covers the western portion of the domain in western Mississippi, Louisiana and Arkansas and the northernmost interior of the domain: Tennessee and West North Carolina.

The available water is highest for the sandy loam, and somewhat lower for the silt loam and loamy sand soil types.

An upland cotton mid-season cultivar was chosen, as representative of the cotton cultivar grown in the southeastern region. We used a May 1st planting date and a November 20th harvest date in the model simulations based on a the USDA table of 'usual planting and harvesting dates by state for U.S. field crops' (http://usda.mannlib.cornell.edu/) and various model sensitivity studies. These dates gave optimal domain-average model yields for present-day climate conditions. Although, this harvest date falls towards the end of the range of harvest dates in this table (the planting date of May 1st falls in the most active planting window in the table), the late harvest date allows for cotton bolls to reach maturity, which did not occur when we used earlier harvest dates in the model simulations.

Cotton model experiments were performed for both dryland and irrigated conditions at all sites. Irrigation occurs when water-stress index levels are less than 0.75 (on a scale of 0 (complete stress) to 1 (no stress)); generic applications of fertiliser were used ($\sim$150 lbs/acre NO_3) to avoid nitrogen stress.

A number of simulations were performed to assess changes in crop yields with the GOSSYM model for the southeastern U.S.A., and for a limited section of the southwestern U.S. (Texas, Arizona, New Mexico, Oklahoma and California). The southeastern U.S. domain covers Louisiana, Arkansas, Mississippi, Alabama, Georgia, South Carolina and Northern Florida and includes only the far West of Tennessee and the East of North Carolina, since cotton is not commonly grown in the Appalachian regions of Tennessee and North Carolina. The present-day climate or baseline experiment was performed using present-day climate for the 36 years (1960–1995) and CO_2 levels of 330 ppm. Note that this baseline or present-day climate will differ from reality in that we have used the same planting date across all grids. Then three sets of simulations were performed with the coarse and fine-scale climate scenarios (36-year length; see Section 2.1) both under dryland and irrigated conditions.

Table VI

Multiple regressions of cotton yield on monthly climate variables for each GCM grid for present-day and for the two climate scenarios for climate change alone – Case (a) for dryland conditions

CSIRO	N	Present day		Coarse-scale		Fine-scale		Coarse-fine scale	
		Variable(s)	R^2	Variable(s)	R^2	Variable(s)	R^2	Variable(s)	R^2
Southwest	49	STR11	0.55	STR7	0.85	STR8	0.83	DSTR8	0.54
coastal		STR11, SOL8	0.63	STR7, SOL10	0.89	STR8, SOL6	0.91	DSTR8, DSOL5	0.71
South-central	56	P11	0.54	STR7	0.68	STR8	0.56	DSOL5	0.57
coastal		P11, SOL5	0.64	STR8, P8(–)	0.87	STR8, P7(–)	0.73	DSTR8, DSOL6	0.72
Centre-west	71	TAV9	0.58	STR8	0.95	STR8	0.88	DSTR8	0.80
inland		TAV6, STR7	0.80	STR8, TAV9	0.97	STR8, TAV11	0.93	DSTR8, DTAV5	0.83
Centre-centre	60	TAV9	0.83	STR10	0.61	STR8	0.61	DSTR8	0.62
inland		TAV9, STR7	0.91	STR10, TAV11	0.71	STR8, P11(–)	0.71	DSTR8, DSTR7(–)	0.69
Centre-east	35	TAV7	0.75	STR7	0.40	P8	0.58	DTAV9(–)	0.51
coastal		TAV7, STR10	0.78	STR7, SOL11	0.64	TAV5, STR10	0.70	DTAV9(–), DSOL10(–)	0.70
Northwest	34	TAV6	0.94	STR8	0.91	STR8	0.70	DSTR8	0.74
inland		TAV6, TAV9	0.95	STR8, P7	0.97	STR8, STR11	0.81	DSTR8, DSOL5(–)	0.80
Northeast	31	TAV9	0.87	P7	0.83	STR11	0.89	DP11	0.56
coastal		TAV8, P5	0.91	STR11, STR8	0.90	STR11, P10(–)	0.95	DP11, DSTR8	0.70

TAV = average daily temperature; STR = water stress index; P = precipitation; SOL = solar radiation; D'VAR" = difference between coarse and fine-scale scenario for variable 'VAR'; 5 = May, . . . , 11 = November. Relationships are positive in the regression equation for the named variables unless otherwise indicated.

Figure 5. CSIRO GCM grid boxes over the Southeastern domain. Masked off region is where cotton is not grown.

For both climate scenarios for the climate change alone case, July and August water stress are dominant influences in most grids, with October and November water stress important in a few other grids. R^2 values for the one-predictor model are generally higher for the coarse scale-scenario, reaching 0.95 in the Center-West grid and 0.91 in the Northwest grid. Only in the Northeast grid in the coarse-scale scenario and in the Center-East grid in the fine-scale scenario is summer precipitation the most important variable determining yield patterns. In both of these grids, water-stress is an important variable with two predictors in the regression equation. As discussed above with the increased temperatures and decreased precipitation in the climate scenarios yields are less sensitive to variations in higher temperatures, nearer to the optimum and are most sensitive to variations in water-stress. In the Southwest grid solar radiation and in the Center-West grid average temperature appears to be important in both scenarios. The correlations of yield with the water stress indicator are consistently positive. Ordinarily one would expect yields to be positively related to precipitation, as well. Surprisingly, however, precipitation has a *negative* coefficient when combined with the water stress indicator in several of the regression models in the different grids. We explored the possibility that the negative relationship with precipitation really indicated a positive relationship through lower cloudiness and greater amounts of solar radiation. This was the case for the south-central grid in the fine-scale but surprisingly not in the coarse-scale scenario. Further investigation will be required to determine the meaning of the negative coefficient. This result is a reminder that multiple regression is only used as a guide to sort out the complex relationships between simulated yields and the plethora of climate variables used by the cotton model.

Overall water-stress levels in conjunction with a more extreme climate (Section 2.1) are higher in the fine-scale scenario except in the Northwest (Figures 4b,c). In the coarse-scale scenario, Central and Southwest Arkansas (covered by the Centre-West and Northwest grids) have the highest water-stress levels (low water-stress index totals) coincident with the highest R^2 values stated above (Figure 4b, Table VI). The highest summer temperature increases and precipitation decreases occur for these grids in the coarse-scale scenario (Figures 1a,c). These grids also have a silt loam soil with lower available water than for the sandy loam soil used in the neighboring central grids (Section 2.3). This combination of more extreme climate change and poorer moisture availability in the soil, may explain the high August water-stress levels simulated (Table VI). These higher water-stress levels also correspond to lower plant weights and larger yield decreases in the coarse-scale scenario in this region compared to the fine-scale scenario (Figures 2a,b). The fine-scale scenario exhibits highest August water-stress levels along the eastern coastal plain (particularly in the North) (Figure 4c). Highest temperature increases and precipitation decreases are found in the fine-scale scenario in this region (>6 °C, Figures 1b,d). As discussed above, in the eastern grids R^2 values are higher in the 1-predictor model in the fine-scale scenario. Over this region plant weights are slightly lower in the fine-scale compared to the coarse-scale scenario. Yields mostly increase in the coarse-scale scenario and decrease in the fine-scale scenario over this region (Figures 2a,b).

The same variables that explain the variability in spatial yield patterns also explain the differences in yields between the coarse and fine-scale scenario. Thus, difference in water stress in August is the dominant variable influencing differences in yield across four of the grid-boxes, in the Southwest, the Centre-West, the Centre-Centre, and the Northwest. It appears as an important variable with two predictors in the regression in these and South-Central and Northeast grid boxes. Difference in solar radiation in spring is important in the South-Centre grid-box in the one-predictor model and is also an important second predictor variable in a number of grid-boxes. Difference in precipitation in November is the most important one-predictor variable in the northeast but difference in November water-stress is almost as important in the one and two-predictor models (not shown). Only in the Centre-East grid box does the difference in August water-stress not appear as an important variable in either the one or two predictor model regression (Table VI). Difference in average temperature between the two climate scenarios seems to be most influential parameter in the Centre-East region. Some of the largest differences in temperatures between the coarse and fine-scale scenarios are predicted over this region (Section 2.1, Figures 1c,d). Summer temperatures are predicted to increase between 4–5 °C in the coarse-scale scenario and up to 7 °C in some parts in the fine-scale scenario (e.g., Figures 1c,d). In addition, this eastern area has sandy loam soils with high drainage rates and low moisture retention capability compared to the neighbouring sandy loam soil types (Section 2.3), thus water-stress is likely to be high and possibly more uniform across the region (Figures 4b,c).

5. Summary and Conclusions

We simulated cotton yields to investigate the effect of the spatial scale of climate scenarios on cotton yields under a number of cases of climate change in the southeastern U.S.A. Under dryland conditions, differences in projected state-level changes in yield simulated with the coarse and fine-scale scenarios were generally found to be significantly different at the 0.05 level for all three cases. Projected yields are generally higher under the coarse-scale scenario in all three cases. When irrigation is applied, differences in yield between the two spatial-scale scenarios both at the state and domain level decrease substantially, showing that through alleviation of water-stress, irrigation may be an effective method of adapting production conditions to changes in climate in the southeastern U.S.A.

For all three cases, the changes in climate hastened phenological processes due to the relatively warmer temperatures. With the change in climate, the number of cotton bolls per plant, which appears to be a strongly related to temperature, increases compared to present-day conditions. Boll weights per plant are lower under the changed climate suggesting that the hastened phenological period does not allow for maximum boll weight development. The higher boll weights with irrigation suggest that water-stress reduces biomass accumulation through photosynthesis under dryland conditions.

Regression analysis was used to examine the most important variables determining spatial patterns for present-day and for the two climate scenarios for the climate change alone case and under dryland conditions. For present-day 1960–1995 climate, spatial variability in summer temperature (during flowering period) is generally the most important variable for determining spatial yield patterns. For the climate change alone case, the two scenarios show water-stress in July and August (flower to open boll period) to be the dominant factor in terms of spatial variability in yields across the domain, with October and November (boll-maturation period) water-stress important in some areas. With the changes in climate, variability in the mean temperatures, which are higher and nearer-to-optimum, does not influence yield variability strongly. However, the higher temperatures, which give rise to higher evapotranspiration rates along with lower precipitation, result in much higher water-stress levels than for present-day conditions, hence variability in water-stress levels becomes the major factor influencing spatial variability in yields. This distinct change in the controlling variables from the present-day to the climate change cases may sound a warning regarding the use of empirical regression models based on current conditions being applied to climate change analysis. Differences in spatial yield patterns between the coarse and fine-scale scenarios are also most strongly influenced by differences in water-stress levels between the two spatial-scale scenarios.

There are many uncertainties in modelling of impacts of climate change on agricultural systems. As discussed earlier (Section 1) there has been much work to examine sources of uncertainties in climate model results that are used to create climate scenarios. We have investigated uncertainty with regard to the effect of two different spatial-scale climate scenarios on cotton yield estimates over the south-eastern U.S.A. In this region, the differences in the spatial scale of the scenario gives rise to two climate scenarios that generally produce significantly different state-level yield changes. It should be noted that the representation of soils in our study is rather coarse even at the 50-km resolution, due to the limited number of soil types with associated parameters available, for use in the GOSSYM model. Additional sources of uncertainty pertaining to the climate scenarios include climate model sensitivity and uncertainties in baseline climate measurements. In addition to climate input uncertainties, there are uncertainties with regard to the crop model. Numerous studies have shown that yield estimates obtained in climate impact assessments vary from one impact model to another (e.g., Mearns et al., 1999). We have not examined this issue here with cotton. As with most impact models, GOSSYM has been developed from datasets from sunlit-controlled environment chambers and is designed to operate at the site-scale. While it has been validated across numerous sites in different regions its applicability across space has not been extensively tested. Additionally, there are numerous management procedures involved in the process of growing cotton, and although many management options are included in GOSSYM, it is not possible to simulate all possible management options that are carried out in the field.

Despite these uncertainties, our study suggests that future climate change with higher temperatures will likely result in higher water-stress levels which may present cotton growers with further water management issues. However, we have shown that despite such large increases in temperature and decreases in precipitation in the climate scenarios used in this study, cotton yields can increase in the future with appropriate management strategies. While we cannot establish that we have greater confidence in the yield results using the higher resolution scenario, we have established that the resolution of climate scenarios is important. Moreover, evidence continues to mount that in many cases higher resolution simulations respond more realistically to high resolution variations in land-surface characteristics (e.g., complex topography, land-cover, and coastlines). We recommend further research into the issue of spatial scale of climate scenarios in other impact sectors, such as water resources, and human health.

Acknowledgements

The authors would like to acknowledge the following staff at the Department of Plant and Soil Sciences at Mississippi State University: Frank Whistler for advice pertaining to soils in the GOSSYM model, Harry Hodges for his helpful comments on an earlier draft of this paper, and Kit Boone for help with model set-up. We also thank Tim Hass at the University of Wisconsin for providing statistical advice. We would also like to acknowledge three anonymous reviewers for their extremely helpful comments which has improved this manuscript. Although the research described in this article has been funded wholly or in part by the United States Environmental Protection Agency through R824997–01–0 to Linda O. Mearns, it has not been subjected to the Agency's required peer and policy review and therefore does not necessarily reflect the views of the Agency, and no official endorsement should be inferred. This research was also funded by the National Aeronautics and Space Administration, MTPE (Grant No. OA99073, W-19, 080).

References

Adams, R. M., McCarl, B. A., and Mearns, L. O.: 2003, 'The Effects of Spatial Scale of Climate Scenarios on Economic Assessments: An Example from U.S. Agriculture', *Clim. Change* **60**, 131–148.

Albers, D. W., Mobley, J., and Tracy, P. W.: 1992, 'Evaluation of GOSSYM-COMAX in Missouri 1987–1991', in *Proc. Beltwide Cotton Conf. National Cotton Council*, Memphis, Tennessee, U.S.A., pp. 1350–1351.

Baker, D. N. and Landivar, J. A.: 1991, 'The Simulation of Plant Development in GOSSYM', in Hodges, T. (ed.), *Predicting Crop Phenology*, CRC Press, Boca Raton, FL, pp. 153–170.

Boone, M. Y. L., Porter, D. O., and McKinion, J. M.: 1993, 'Calibration of GOSSYM: Theory and Practice', *Comput. Electron. Agric.* **9**, 193–203.

Brown, R. A. and Rosenberg, N. J.: 1999, 'Climate Change Impacts on the Potential Productivity of Corn and Winter Wheat in their Primary United States Growing Regions', *Clim. Change* **41**, 73–107.

Carbone, G , Kiechle, W., Locke, C., Mearns, L. O., McDaniel, L., and Downton, M.: 2003, 'Response of Soybeans and Sorghum to Varying Spatial Scales of Climate Change Scenarios in the Southeastern United States', *Clim. Change* **60**, 73–98.

Doherty, R. and Mearns, L.: 1999, *A Comparison of Simulations of Current Climate from Two Coupled Atmosphere-Ocean Global Climate Models against Observations and Evaluation of their Future Climates*, Report to the National Institute for Global Environmental Change (NIGEC), Boulder CO, ESIG/NCAR.

Dudek, D. J.: 1989, 'Climate Change Impacts upon Agriculture and Resources: A Case Study of California', in Smith, J. and Tirpak, D. (eds.), *The Potential Effects of Global Climate Change on the United States, Appendix C: Agriculture, Vol. 1*, EPA-230-05-89-053.

Dugas, W. A., Heuer, M. L., Hunsaker, D., Kimball, B. A., Lewin, K. F., Nagy, J., and Johnson, M.: 1994, 'Sap Flow Measurements of Transpiration from Cotton Grown under Ambient and Enriched CO_2 Concentrations', *Agric. For. Meteorol.* **70**, 231–246.

Easterling, W., Chhetri, N., and Niu, X.: 2003, 'Improving the Realism of Modeling Agronomic Adaptation to Climatic Change: Simulating Technological Substitution', *Clim. Change* **60**, 149–173.

Felzer, B. and Heard, P.: 1999, 'Precipitation Differences amongst GCMs Used for the U. S. National Assessment', *J. of the American Water Resources Association* **35**, 1327–1339.

Giorgi, F., Hewittson, B., Christensen. J., Hulme, M., Von Storch, H., Whetton, P., Jones, R., Mearns, L., and Fu, C.: 2001, 'Regional Climate Information – Evaluation and Projections', in Houghton, J. T., Ding, Y., Griggs, D. J., Noguer, M., van der Linden, P. J., Dai, X., Maskell, K., and Johnson, C. A. (eds.), *Climate Change 2001: The Scientific Basis*, Chapter 10, Cambridge University Press, Cambridge, U.K., pp. 583–638.

THE EFFECTS OF SPATIAL SCALE OF CLIMATE SCENARIOS ON ECONOMIC ASSESSMENTS: AN EXAMPLE FROM U.S. AGRICULTURE

RICHARD M. ADAMS [1]*, BRUCE A. McCARL [2]* and LINDA O. MEARNS [3]

[1]*Department of Agricultural and Resource Economics, Oregon State University, Corvallis, OR, U.S.A.*
E-mail: Richard.M.Adams@oregonstate.edu

[2]*Department of Agricultural Economics, Texas A&M University, College Station, TX, U.S.A.*
E-mail: mccarl@tamu.edu

[3]*Environmental and Societal Impacts Group, National Center for Atmospheric Research, Boulder, CO, U.S.A.*
E-mail: lindam@ucar.edu

Abstract. The appropriate level of spatial resolution for climate scenarios is a key uncertainty in climate impact studies and regional integrated assessments. To the extent that such uncertainty may affect the magnitude of economic estimates of climate change, it has implications for the public policy debates concerning the efficiency of CO_2 control options. In this article, we investigate the effects that different climate scenario resolutions have on economic estimates of the impacts of future climate change on agriculture in the United States. These results are derived via a set of procedures and an analytical model that has been used previously in climate change assessments. The results demonstrate that the spatial scale of climate scenarios affects the estimates of both regional changes in crop yields and the economic impact on the agricultural sector as a whole. An assessment based on the finer scale climatological information consistently yielded a less favorable assessment of the implications of climate change. Regional indicators of economic activity were of opposite sign in some regions, based on the scenario scale. Such differences in economic magnitudes or signs are potentially important in examining whether past climate change assessments may misstate the economic consequences of such changes. The results reported here suggest that refinement of the spatial scale of scenarios should be carefully considered in future impacts research.

1. Introduction

The coarse spatial scale of climate scenarios generated from global general circulation models (GCMs) – on the order of hundreds of kilometers – has long been a concern to those charged with assessing the consequences of climate change (Gates, 1985). This concern arises because of the mismatch between the scale of the climate change information from GCMs and the resolution of most climate effects models. Today, climate model procedures and other techniques are available to create results on a finer regional scale (Giorgi and Mearns, 1999; McGregor, 1997; Wilby and Wigley, 1997). Thus, an important question in ongoing and future

* Seniority of authorship is shared.

Climatic Change **60:** 131–148, 2003.

research on the effects of climate change involves the appropriate level of spatial resolution for climate scenarios and more to the point, the gain in information content from use of a finer scale than offered by the global GCMs.

Readers should note that a scale investigation on climate information addresses only a part of the total scale issue. In particular one could consider the effects of alterations in the resolution of the economic and crop models, as well as the climate models. Certain aspects of such an investigation have been pursued elsewhere. For example, information on the number of and categorization of representative regions in economic models is reviewed from a production response viewpoint in Just and Pope (2001); a climate change perspective in Antle (1996); an aggregate firm representation viewpoint in Onal and McCarl (1991); and a crop model 'scaling up' viewpoint in Hansen and Jones (2000). In this paper we confine ourselves to an examination of the consequences of changes in climate scenario regional specification, holding the scaling characteristics of the crop and economic models constant.

In an agricultural context, it is expected that the use of finer scale regional climate information will affect climate change appraisals in several ways:

1. Finer scale data by their nature will more adequately represent localized variation in climate and match better with physically based crop models and localized knowledge (soils, proximity to water, elevation, etc.).
2. The resultant yield estimates will be available on a finer scale.
3. Data for aggregate economic models developed on regional average yields are more likely to match the underlying data being developed on the basis of regional average climate change effects. This would contrast with current practices where either a single simulation location is used to represent a region and is subjected to a single grid cell or spatially interpolated climatic outcome; or coarse scale climate model results are combined with a fine resolution gridded observed data set, which merely gives the illusion of a fine scale treatment.
4. Adaptation possibilities can be more carefully tailored to local realizations of broad climatic outcomes.

Given these possible effects, the underlying analytical issue involves whether the use of finer scale climate information in climate change assessments will lead to different impact estimates relative to assessments done based on larger scale data. This paper addresses this question from an economic viewpoint, examining whether the change in scale of the climate information alters the results of the economic component of an integrated climate change assessment. Contrasts in both the changes in mean and variability of yields with spatial scale are considered.

2. Study Procedures and Yield Simulation Results

The economic assessment performed here builds on information from climatologists, crop modelers and geographers, as illustrated by the earlier papers in this volume. These inputs include alternative climate scenarios developed by climatologists using global and regional climate models, along with crop yield and crop irrigation water requirement sensitivity results developed by employing crop simulation models or other climate-sensitive estimation techniques. An overview of the procedures used to draw together the inter-disciplinary information and the basic yield results appears below as well as in the introductory editorial of this volume (Mearns, 2003). Moreover, an overview paper (Mearns et al., 2003a) is in press. More complete discussions relative to the results from other team members are found in other articles of this issue of *Climatic Change*.

2.1. CLIMATE SCENARIOS

The study required linked GCM and regionalized finer scale climate scenarios. The regional climate model RegCM (Giorgi et al., 1993a, b), run at a 50 km grid point spacing, was used to generate relatively high resolution scenarios as described in Mearns et al. (2003b). The RegCM was nested within the Australian Commonwealth Scientific and Industrial Research Organization (CSIRO) GCM (Watterson et al., 1997) control and doubled CO_2 experiments. The CSIRO GCM uses a grid corresponding to about 5° in physical space, roughly 400 km on a grid side. In the nesting procedure, the large scale response to changes in climate forcings is provided by the coarse resolution CSIRO boundary conditions, but the mesoscale response (sub-GCM grid scale) is provided by the RegCM (Giorgi and Mearns, 1999). RegCM was run for three regions using the same series of years from the CSIRO for boundary conditions: (1) the western two-thirds of the conterminous United States (Giorgi et al., 1998) (2) the Great Lakes (Bates et al., 1995 and unpublished data) and (3) most recently, the Southeast (Mearns et al., 2003b). In any given region, the CSIRO and RegCM climate change projections are similar for very broad averages, but the finer resolution information shows substantial differences within regions, especially for precipitation (Giorgi et al., 1998; Mearns et al., 2003b).

From these runs (CSIRO and RegCM) two fully regionalized 50 km grid climate change scenarios to be used as input to the crop models were formed by applying the projected ratios ($2 \times CO_2$/control) or monthly mean climate differences from each climate model for maximum and minimum temperature, precipitation, and incident solar radiation to daily meteorological observations (1961–1985). The results for the CSIRO scenario used the same changes for each 50 km grid point contained within the 400 km grid cells. For the RegCM fine resolution scenario, each 50 km grid received a unique set of changes. See Mearns et al. (2003b) for more details on how the scenarios were formed.

2.2. PROJECTED CROP YIELD AND WATER USE EFFECTS OF CLIMATE CHANGE

The economic analysis does not use the climate data directly. Rather, projections of crop yield and irrigation water use associated with those climate data were needed. To obtain such estimates, the CSIRO and RegCM regionalized scenarios were used as the weather input to crop simulators. The crop simulators employed for corn, rice, sorghum, soybean, and wheat were CERES and CROPGRO (Tsuji et al., 1994) while the GOSSYM model (Hodges et al., 1998; Reddy et al., 1997) was used for cotton. These models were run on each of the relevant cells in the 50 km grid as described in the Carbone et al., Doherty et al., and Tsvetsinskaya et al. (2003).

Crop model runs for both dryland and irrigated conditions were performed for a base case (using observed climate and a CO_2 level of 330 ppm), an elevated CO_2 case (540 ppm), as well as elevated CO_2 with farmer adaptations. Adaptations included adjustments in planting dates and/or changes in cultivars. It should be noted that the adaptations were applied in a manner to optimize the response of the crop yields to the new climates. The adaptations assume 'clairvoyance' or perfect knowledge on the part of the farmer, i.e., that climate changes are known completely in advance, and that perfect adaptations are instantaneously adopted. In this regard, such adaptations are somewhat unrealistic. While the non-adaptation cases are also unrealistic, it should be understood that true or actual adaptations would likely fall somewhere between the 'clairvoyant' adaptation results and the non-adaptation results. More realistic handling of adaptations is discussed in Schneider et al. (2000), and more completely in Easterling et al. (2003). Because we feel the 'reality' lies somewhere between the two cases, we report the results for both cases.

For input to the economic model, state average yields, weighted according to the state average acreage (based on 1997 acreage allocations), for the base case and the different scenarios were calculated. The alternative levels of regional resolution or scale of the climate scenarios resulted in substantial differences in estimated yield sensitivity for most crops across the cases. Selected results are presented in Table I, which provides national average percentage changes in yields and water use, by scenario/case.

The general pattern is for yields to be lower under the finer scale of resolution, showing that the subregional variations in climate did not match up as favorably with the physical resource characteristics as did the coarser scale climate change. For example, in a region characterized by a large elevation change, the warmer conditions and reduced precipitation may have fallen more severely on low lying, productive areas reducing yield more than would happen under the broad regional average climate change. Carbone et al., Doherty et al., and Tsvetsinskaya et al. (2003), present detailed discussion of the causes for the contrasts in changes in crop yield based on the scale of the climate scenarios, while Mearns et al. (1999, 2001b) discusses these effects in the Great Plains region.

Table I
National average percentage change in yield, yield variability and water use

Crop	Irrig. status	Effect	CSIRO		RegCM	
			Without adaptation	With adaptation	Without adaptation	With adaptation
Corn	Dryland	Yld	0	11	–6	7
		Ycv	–26	–24	–17	–22
	Irrigated	Yld	–16	–11	–15	–10
		Ycv	–16	–12	–17	–14
		Wuse	–24	–27	–6	–7
Cotton	Dryland	Yld	25	48	22	41
		Ycv	–16	–14	–14	–12
	Irrigated	Yld	46	54	45	52
		Ycv	–62	–61	–58	–59
		Wuse	57	69	63	77
Soybeans	Dryland	Yld	33	42	5	17
		Ycv	33	42	34	44
	Irrigated	Yld	42	43	42	44
		Ycv	29	34	48	38
		Wuse	20	20	49	49
Hard red spring wheat	Dryland	Yld	–11	6	–2	18
		Ycv	–25	–24	–33	–33
	Irrigated	Yld	–19	–19	–12	–11
		Ycv	–7	–6	–1	–2
		Wuse	–25	–24	–18	–22
Hard red winter wheat	Dryland	Yld	–2	2	4	7
		Ycv	–11	–17	–16	–26
	Irrigated	Yld	5	5	0	0
		Ycv	23	23	12	12
		Wuse	–15	7	–24	–7
Soft white wheat	Dryland	Yld	–9	–8	–5	–4
		Ycv	5	5	13	13
	Irrigated	Yld	–12	–12	–5	–5
		Ycv	21	21	20	5
		Wuse	–48	–32	–29	–11
Durham wheat	Dryland	Yld	41	47	22	32
		Ycv	73	63	33	21
	Irrigated	Yld	23	23	20	20
		Ycv	25	26	27	27
		Wuse	–20	–20	–11	–11
Sorghum	Dryland	Yld	–7	4	–6	5
		Ycv	–18	–16	–21	–17
	Irrigated	Yld	0	2	–1	3
		Ycv	6	11	8	15
		Wuse	8	8	8	8

Notes: Yld is the yield effect. Ycv is yield coefficient of variation. Wuse is irrigation water use.

Dryland corn 'without adaptation' can be used to illustrate the lower regionalized yields for the finer scale. Here the RegCM run shows a 6% yield loss as opposed to no change under CSIRO, and for the 'with adaptation' case, RegCM yields increase by 7%, while CSIRO increases by 11%.

Underlying these aggregate data are substantial regional differences. Table II presents results for selected regions relevant to the scale of economic modeling (Section 2.3). In the Corn Belt states, without adaptation yields generally decreased across both levels of scale, but the decreases were significantly larger with the fine scale scenario. With adaptations, the coarse scale exhibits small increases whereas the fine scale results remain negative. This same pattern is seen in sorghum yields. Contrasts in irrigated yields are always smaller across the spatial scales. Decreases in corn were also obtained in the southeast (see Tsvetsinskaya et al., 2003, for details) for both scenarios, but particularly in the Carolinas, decreases were larger for the high resolution scenario. This was not necessarily true for results in all regions. In the Northern Plains region, for example, corn yields with no adaptation increased by about 22% for the RegCM scenario but only 14% for the coarse resolution CSIRO scenario. In general, the areas where the RegCM scenario produced higher yield results for many crops (with and without adaptations) were in the North Plains, South Plains, and Delta States. The CSIRO scenario generally produced higher yields in the Corn Belt, Southeast, North East, and Appalachia. In other regions (e.g., Mountain, Pacific, and Lake States) results were more dependent on the particular crop.

The results also show that cotton faired better than most other crops. In the Delta States, the Southern Plains, and Southeast with adaptations, cotton yields increased substantially for both scenarios under dryland and irrigated conditions, but the coarse scenarios produced larger increases.

The pattern of water use for irrigated crops is consistent across both spatial scales (Table I), although there are differences in magnitudes and signs across crops. Water use for irrigated crops tends to increase (decrease) under conditions where irrigated yields increase (decrease). The decrease in water use reflects reduced water requirements where there is a decrease in actual growing period, which reduces the production of biomass; this effect dominates over increased water need due to higher potential evapotranspiration resulting from increased temperatures with climate change. At the national scale of aggregation, of course, these relationships are only approximate. In the economic model, water use is important since changes in water use translate into changes in costs of production, in addition to the effects on revenues associated with changes in yields. Production costs change because water is an input that must be obtained through pumping or other sources.

2.3. ECONOMIC MODELING

The above changes in crop yield, while revealing information on regional crop sensitivity, do not give a full measure of changes in aggregate crop production

Table II
Percent change in crop yields from base yield for selected regions

Region	Crop	Without adaptation				With adaptation			
		CSIRO		RegCM		CSIRO		RegCM	
		Dry.	Irr.	Dry.	Irr.	Dry.	Irr.	Dry.	Irr.
Lake	Corn	3	–12	0	–9	9	–7	4	–4
States	Soybeans	106	56	65	64	116	63	79	70
	HRSW	–20	–24	–20	–19	–4	–24	–3	–18
	HRWW	3	1	12	12	3	1	12	12
	SOFT	20	12	30	45	22	13	35	46
Corn	Cotton	–2	37	–19	37	29	48	15	48
Belt	Corn	–5	–20	–17	–18	5	–13	–9	–13
	Soybeans	33	39	–3	41	37	42	2	45
	SOFT	–7	–8	–2	–2	–6	–7	–1	–1
	Sorghum	–4	–11	–19	–11	3	–11	–12	–10
	Rice	N/A	–6	N/A	–14	N/A	2	N/A	4
Northern	Corn	14	–16	22	–14	27	–14	48	–8
Plains	Soybeans	58	52	62	54	58	41	65	41
	HRSW	–19	–19	–13	–10	–1	–18	4	–9
	HRWW	3	–2	5	–6	6	–2	9	–5
	DURW	37	23	20	20	44	23	30	20
	Sorghum	–9	3	4	3	1	4	11	4
Delta	Cotton	–2	21	–9	19	24	31	15	28
States	Corn	–2	–12	4	–9	9	–8	11	–6
	Soybeans	–37	24	–44	14	1	28	2	25
	SOFT	–23	–24	–16	–18	–23	–24	–16	–18
	Sorghum	–37	–13	–35	–15	–21	–10	–20	–8
	Rice	N/A	–3	N/A	–7	N/A	1	N/A	6
Southern	Cotton	57	78	67	76	85	82	84	80
Plains	Corn	13	–15	34	–13	60	–12	81	–12
	Soybeans	–38	8	6	8	51	14	51	8
	HRWW	–13	–10	7	–13	–8	–10	11	–13
	Sorghum	–3	–6	–11	–6	11	–3	5	–2
	Rice	N/A	0	N/A	1	N/A	8	N/A	14

Lake States: Minnesota, Wisconsin, Michigan.
Corn Belt: Iowa, Missouri, Illinois, Indiana, Ohio.
Northern Plains: North Dakota, South Dakota, Nebraska, Kansas.
Delta States: Arkansas, Louisiana, Mississippi.
Southern Plains: Texas, Oklahoma.
NA = not applicable; Dry. = dryland; Irr. = irrigated.
Other regions not shown include the Southeast, Pacific, Appalachia, Northeast, and Mountain States.

(regionally or nationally) nor of the sensitivity of economic welfare at the national level. Changes in crop mixes across regions and altered market prices can partially offset the crop yield effects. Consequently, it is important to derive sector wide information on the sensitivity of economic outcomes to information differences based on spatial scale. This requires use of economic procedures which account for producer and consumer adaptations to changes in yields and prices.

To measure the economic effects, we used an economic model of the U.S. agricultural sector called the Agricultural Sector Model (ASM). ASM was originally developed in a cooperative agreement with USDA in a thesis by Baumes (1978). It was subsequently improved for pesticide analysis by Burton (1982) as reported in Burton and Martin (1987) and then documented by Chattin et al. (1978) and later by Adams et al. (1984) and Chang et al. (1992) with components discussed in McCarl and Spreen (1980), McCarl (1982), and Hamilton et al. (1985). In the late 1970s it began to be used for environmental and resource analysis, first on biofuels (Tyner et al., 1979; Chattin, 1982; Hickenbotham, 1987), then on ozone (Hamilton, 1985; Adams et al., 1984), acid rain (Adams et al., 1993), soil conservation policy (Chang et al., 1994), global climate change (Adams et al., 1988, 1990, 1995b, 1999, 2001; McCarl, 1999; Reilly et al., 2000, 2002), and climate change mitigation (Adams et al., 1993; McCarl and Schneider, 2001). It has also been used extensively for research evaluation (Tanyeri-Abur, 1990; Coble et al., 1992; Chang et al., 1991). A variant of the model has been adapted and used for about 20 years by the USDA Economic Research Service (House, 1987; Heimlich et al., 1998).

This study also employs a stochastic variant of ASM, known as SPRASM, as documented in Lambert et al. (1995) and as used in an ENSO setting by Adams et al. (1995a); Solow et al. (1998); Chen and McCarl (2000) and in a climate change setting by Chen et al. (2001).

ASM is a spatially disaggregate model that simulates the economic equilibrium that arises in the U.S. agricultural sector. It considers the regional impacts of yield changes, with endogenous price adjustments (McCarl and Spreen, 1982; McCarl et al., 2000). ASM represents production and consumption of primary agricultural products including both crop and livestock products.

The primary attribute of SPRASM over the deterministic ASM is that it allows yield variability to enter through states of nature dependent on crop yields. Thus, the SPRASM model allows assessment not only of the expected impacts of climate change, but also price, production, and regional impacts under stochastic crop yield outcomes. Stochastic yield data for the base case (before climate change) were derived as the residuals from a set of trend models estimated from state level yields as reported in the annual USDA Agricultural Statistics volume following Thaysen for each crop region using 1977 to 1999 observations. Residuals were added to the ASM base year observations to give 23 equally likely yield outcomes resulting from weather or other stochastic factors. Use of the discrete, empirical yield distributions, derived from the observed year-to-year yield outcomes, preserves both interregional and intercrop correlation.

The economic welfare of participants in agriculture is measured in both ASM and SPRASM as the sum of consumers' and producers' surplus. These are monetary measures that capture, through supply and demand curves, the effects of changes in production and demand on producers' profits and consumers' (both domestic and foreign) expenditures. In the aggregate, these measures indicate potential gains or losses in societal well-being. Individually, they provide information on how the gains or losses are distributed.

The basic ASM economic model contains 63 primary spatial units of analysis, which correspond to states in most cases except in the Midwest, Texas, and California, where sub-state units are delineated. These units are aggregated into ten mega-regions (such as in Table II) for some calculations, including those generated by SPRASM. Specifically, SPRASM was formed as a weighted average of all the budgetary and crop mix data in ASM. These data were aggregated to the ten USDA agricultural production regions, with the same demand relationships (curves) facing each state of nature, but actual processing and consumption varying in response to market conditions and prices. In addition, SPRASM incorporates state-of-nature dependent withdrawals and additions to stocks of each commodity. To account for stock changes, an equation was added for each commodity that equates expected stock additions and withdrawals.

This assessment builds upon a number of previous efforts, the most recent of which are the U.S. National Assessment (Reilly et al., 2000, 2002; McCarl, 1999) and an assessment sponsored by the Electric Power Research Institute (reported in Adams et al., 1999). The other notable direct ancestor of this work is Adams et al. (1990). The modeling framework allows for eight principal effects of climate change to be considered. These involve the effects of climate change on

1. crop yields,
2. irrigated crop water use,
3. irrigation water supply,
4. livestock performance,
5. livestock pasture usage,
6. animal unit month grazing supply,
7. international trade,
8. pesticide usage, and
9. yield coefficient of variation (stochastic version only).

As detailed in Adams et al. (1999), the model also includes estimates of the indirect influence of climate change on livestock and crop yields. These indirect effects arise through feedbacks between crop yields and farmers' adjustments in production practices and input usage. Most of the crop yield and irrigation water use response data used in the present assessment were developed in this project (see the earlier articles in this special issue for a complete listing of crops treated in

Figure 1. Study area and grid.

Among factors simulated by EPIC are evapotranspiration (based on the Penman–Monteith model), soil temperature, crop potential growth, growth constraints (water stress, stress due to high or low temperature, nitrogen and phosphorus stress) and yield. EPIC uses a single model for simulating all crops, although of course, each crop has unique values for model parameters. The crop growth model relies on radiation-use efficiency to calculate photosynthetic production of biomass. The potential biomass is adjusted daily for stress from the following factors: water, temperature, nutrients (nitrogen and phosphorus), aeration and radiation.

Atmospheric CO_2 concentrations in EPIC influence photosynthesis by modification of the radiation-use efficiency term. They also influence water use efficiency by modification of the stomatal conductance term in the equation of evapotranspiration. Crop-specific functions were developed by solving response curves of radiation-use efficiency versus CO_2 from the literature. We define water use efficiency as the ratio of crop yield to actual evapotranspiration (Easterling et al., 1992b). The CO_2 fertilization algorithm in EPIC was modified by Stockle et al. (1992) to account for the interactive effects of vapor pressure deficit (VPD) and increased CO_2 on radiation-use efficiency. The higher the VPD, the lower is the positive effect of CO_2.

For this study CO_2 concentrations are initialized at 330 ppm (baseline ambient level) and then allowed to increase in intervals of 17.5 ppm with each step of climate change (discussed below) to a final concentration of 540 ppm. Crop yields are estimated by multiplying the above-ground biomass at maturity (determined by accumulation of heat units or specified harvest date) by a harvest index (crop yield divided by above ground biomass) for the particular crop.

EPIC requires data on soil physical properties (for example, bulk density, water-holding capacity, wilting point) and management (for example, fertilization, tillage, planting, harvesting, irrigation). The weather variables necessary for driving the EPIC model are daily values of precipitation, minimum/maximum air temperature, solar radiation, windspeed and relative humidity. A detailed discussion of EPIC input data attributes unique to this study comes in the next subsection.

EPIC has been subjected to numerous validation exercises. Extensive tests of EPIC simulations were conducted at over 150 sites and on more than 10 crop species and generally those tests concluded that EPIC adequately simulated crop yields (Kiniry et al., 1990; Rosenberg et al., 1992). Easterling et al. (1996) specifically validated the EPIC yield predictions during recent episodes of climate extremes that emulate expected attributes of climate warming in the Great Plains.

3.2. EPIC INPUT DATA FOR THIS STUDY

3.2.1. *Soil Data*

The State Soil Geographic (STATSGO) database (USDA, 1994) was used with a geographic information system to determine the *dominant* soil (defined as the most extensive soil) for each of the different grid scales. The STATSGO soil association maps are compiled from generalized soil survey maps at 1:250,000 scale. The STATSGO data are divided into map units with a minimum resolution of 625 hectares, which can contain up to 21 soils. The percentage of each soil is known within each map unit but its distribution within each map unit is not known. Therefore the *dominant* soil is only an estimate since a map unit is not necessarily totally contained within a grid cell. STATSGO also lists a prime farmland classification for each soil. This classification rates farmland from 0 to 9 based on soil and climatic properties and was used to determine the *dominant* soil type depending on what crop was being modeled in an area. Only soils with the rating of prime farmland were considered for maize, soybeans, and wheat.

3.2.2. *Climate Data and Climate Change Scenario*

The baseline climate was compiled from 1960–1995 daily observations of maximum/minimum temperature, precipitation, radiation, relative humidity and windspeed (described in Mearns et al., 2003, this issue). The climate change scenario of the RegCM limited area climate model is described more fully in Mearns et al. (2003). The larger project that incorporates this research developed climate change scenarios at both low resolution (general circulation model) and high-resolution (limited area model nested within a general circulation model) scales. The rationale for the climate models used in this research appears in Mearns et al. (2003). Because this research does not examine issues of spatial scale directly, we elected to use the high-resolution scenario only. This decision was also supported by previous research that concluded that the additional spatial detail of high-resolution climate change scenarios adds regionally-detailed information to adaptation analysis not available with the use of low-resolution scenarios (Easterling et al., 2001).

A major challenge in constructing our methodology was the development of a time-evolving climate change scenario derived of an equivalent doubling of atmospheric CO_2 equilibrium solution. Our test of logistic adaptation required a time-evolving climate change scenario. By dividing the RegCM $2 \times CO_2$ climate change into 12 equal fractions we created a 'mock' transient (time-evolving) climate change scenario. Each of the 1/12 segments of climate change (temperature, precipitation, relative humidity, windspeed, solar radiation) was used to adjust the 1960–1995 baseline daily climate following standard procedures (described fully in Mearns et al., 2003, this issue). While we recognize that changes in climate variability caused by mean changes in climatic elements using this scheme are crucial to agricultural production, we do not analyze the effects of change in variability on adaptation of crops *per se* in this paper. This is a recognized deficiency of this research and is being addressed explicitly in follow on research.

The projection of future rates of increase in atmospheric CO_2 and resulting climate change is a matter of speculation and debate. For purposes of discussion only, we assume that it takes 60 years to reach the $2 \times CO_2$ climate change. Hence, each of the 12 fractions represents 5 years of climate change. To simplify reporting of the results of our crop modeling, we convert the resulting 12 5-year yield simulations into three 20-year means (1/3, 2/3, 3/3). Shorter intervals than 20 years do not allow adequate time for distinguishing the dynamic effect of climate change on adaptation. The obvious lack of realism in this approach is compounded by an implicit assumption of a linear trajectory of climate change.

The RegCM climate changes (temperature and precipitation) are summarized for the corn growing season (April–August) in a set of maps (Figures 2a–c, 3a–c). The changes varied widely across the Southeast (Figures 2a–c, 3a–c). In general, the southwestern portions of the region became slightly warmer and much wetter. The northeastern portions became much warmer and much drier. A positive temperature gradient extended generally from the less warm southwestern quadrant to the much warmer northeastern quadrant of the region. In the first 20 years of change (Figure 2a), the warming was greatest along the eastern seaboard where the temperature increase of several grid boxes averaged as much as 2.1 °C. A small, isolated area of large temperature increase is discernable in southern Louisiana. Most of the area of the Gulf States and the Appalachian Mountains showed little or no warming in this early period. The spatial distributions of warming generally were maintained throughout the latter two thirds of the climate change, with the warming intensifying everywhere (Figures 2b,c). By the third 20 years of climate change, large parts of the eastern seaboard states had warmed by a average of up to 6.3 °C while much of the southwestern part of the area had warmed little (<2.0 °C) if any.

The spatial distribution of RegCM precipitation changes was the mirror opposite of the temperature change distribution, with the largest decreases in all three periods distributed across the northeast quadrant of the region and the largest increases distributed across the southwest quadrant (Figures 3a–c). The areas of

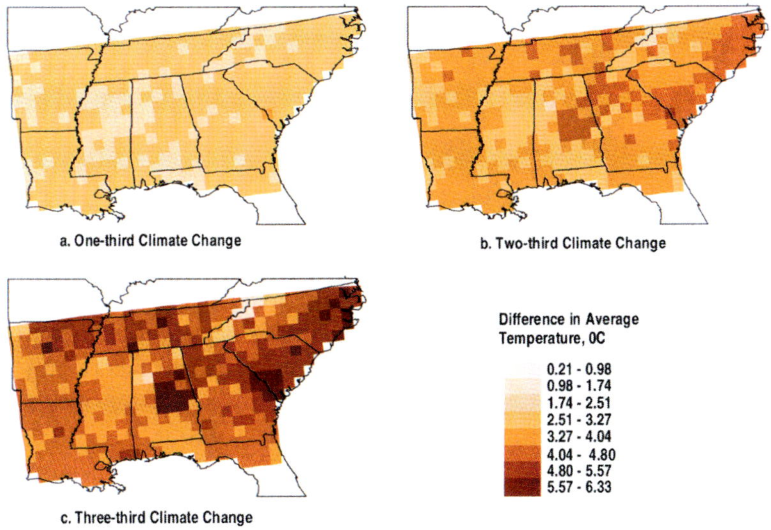

Figure 2. Change in temperature (°C) relative to 1960–1995 baseline climate simulated by the RegCM model. (a) 1/3 of 2 × CO_2 climate. (b) 2/3 of 2 × CO_2 climate. (c) 3/3 of 2 × CO_2 climate (or 'equivalent doubling CO_2 equilibrium').

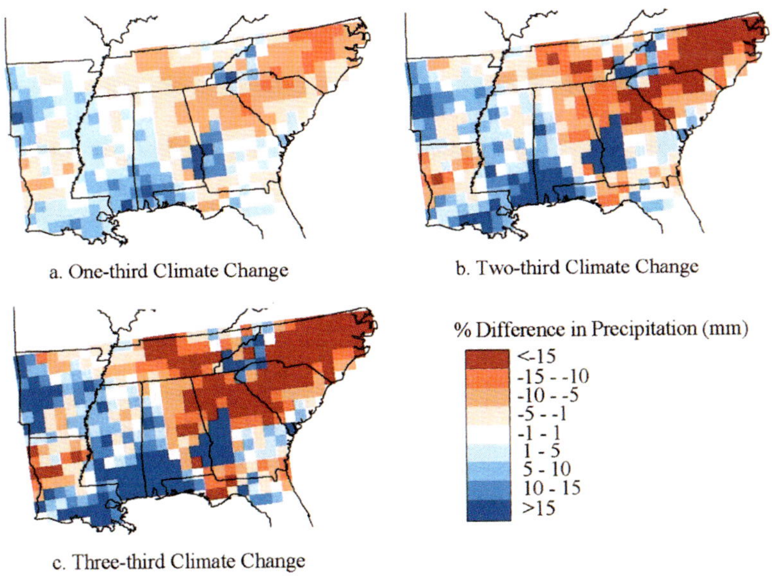

Figure 3. Change in precipitation (%) relative to 1960–1995 baseline climate simulated by the RegCM model. (a) 1/3 of 2 × CO_2 climate. (b) 2/3 of 2 × CO_2 climate. (c) 3/3 of 2 × CO_2 climate (or 'equivalent doubling CO_2 equilibrium').

largest decrease in precipitation lost more than 10% of total growing season rainfall while the areas of largest increase gained more than 15% of total growing season rainfall.

3.2.3. *Preparation of EPIC Representative Farms*

Representative EPIC farms developed by Brown and Rosenberg (1999) for major watersheds of the United States were adapted to the study region. Detailed profiles of relevant production characteristics (cultural practices, planting and harvesting dates, heat units required for maturity) were compiled for a selection of 'typical' farm enterprises across southeast. Although EPIC is capable of simulating many of the crop species grown in the southeastern U.S., we focus on corn only because it is the most extensively grown crop in the region, it is highly sensitive to climate stress, and it is responsive to agronomic management. For the purpose of testing our methodology, one crop is adequate. However, we recommend that other crops, some in rotation, be tested in subsequent research. Representative farms data were gridded such that a single farm enterprise growing corn was positioned at the centroid of each grid box.

3.3. PROCEDURES

Four different adaptation scenarios (including no-adaptation) were constructed to capture the range of possibilities of farmer response to climate change in the EPIC model. They include: (1) *clairvoyant*; (2) *lagged*; (3) *logistic*; and (4) *no-adaptation*.

3.3.1. *Clairvoyant Adaptation*

As noted above, clairvoyant adaptation occurs when all farmers instantly recognize and correctly respond to climate change. EPIC was modified to represent clairvoyant adaptation by optimizing planting dates and cultivar growing season lengths in synchrony with the evolving climate change. The results are interpreted as the 'best case' of adaptation and consonant with the majority of adaptation modeling studies in agriculture.

3.3.2. *Lagged Adaptation*

Lagged adaptation occurs initially when all farmers fail to recognize climate change until 10 years (two 5-year increments of climate change) have elapsed and then they all adapt in unison optimally with respect to the average climate of those 10 elapsed years. From that point on, the farmers remain one full 10-year step behind the current 5-year increment of climate change as they traverse the entire 60 years of climate change. Thus, farmers are assumed to behave as though they fail to recognize the climate change or refuse to believe that the change warrants action for 10 years. EPIC was modified to represent lagged adaptation by optimizing planting dates and cultivar growing season lengths to the average climate of the 10 years prior to the current 5-year increment of climate change. We recognize that few if any farmers are likely to experience 10 years of steady climate change without adapting. However, we deliberately chose a lag that is at the outer bound of failure to act. Clearly, the results are to be interpreted as the 'worst case' of adaptation (save no adaptation at all) and, together with the clairvoyant case, bound the envelope of adaptation.

3.3.3. *Logistic Adaptation*

The logistic curve (Equation (1)) is used to allocate the fraction of farmers, $f(t)$, who adopt changes in planting dates and varietal strains at time t relative to the onset of a 5-year segment of climate change. This assumption stipulates that a small fraction of farmers adapt immediately to the first and each subsequent 5-year segment of climate change. Those early adaptors (f_1) are followed by a larger subgroup of early middle adopters (f_2) that delays adaptation for 10 years and remain 10 years behind in each of the subsequent 5-year segments of climate change. Subgroup f_2 is followed by another large subgroup of late middle adopters (f_3) that delays adaptation for 15 years and remains 15 years behind. Finally, a small subgroup of laggards (f_4) delays adaptation for 20 years (i.e., it begins adaptation at the beginning of the fourth 5-year segment of climate change) and remains 20 years behind. Hence, it takes the first 20 years of climate change for all farmers to begin to adapt. Thereafter, all farmers adjust with each new 5-year segment of climate change, but at the lag times f_1–f_4 specified above. The proportion of farmers allocated among f_1–f_4 in the hypothetical fifth 5-year time slice (years 25–30) is shown in Figure 4 according to Equations (2)–(5) below.

We recognize that issue can be taken with the length of time assumed for cultivar adaptation to be fully adopted (20 years); however, the literature suggests large variation in times required for full adoption of agricultural innovations ranging from 10 years for fertilizer application to 30 years for the full adoption of a new crop species such as soybeans (Reilly, 1995). By lagging 20 years behind the current climate changes, the latter subgroup f_4 consists essentially of non-adaptors who misread the signal of climate change for a variety of reasons ranging from poor knowledge to rigid adherence to tradition. Although we do not explicitly account for farm failure, we presume that this latter group would have a large attrition rate due to inefficient response to climate change. However, it is reasonable to expect that there will be replacement within this subgroup; moreover, we assert that each subgroup will be replenished with new generations of farmers through time, thus maintaining the logistic form of adaptation.

This procedure causes adaptation to begin slowly with relatively few farmers adapting, followed by rapidly increasing numbers adapting less efficiently until the rate slows as full saturation of adaptation at varying degrees of efficiency is achieved by the end of the 20-year cycle.* This sequence is explicit in Figure 5,

* Or 1/3 of the climate change that is expected from an equivalent doubling of atmospheric CO_2, which is assumed to be reached in 60 years.

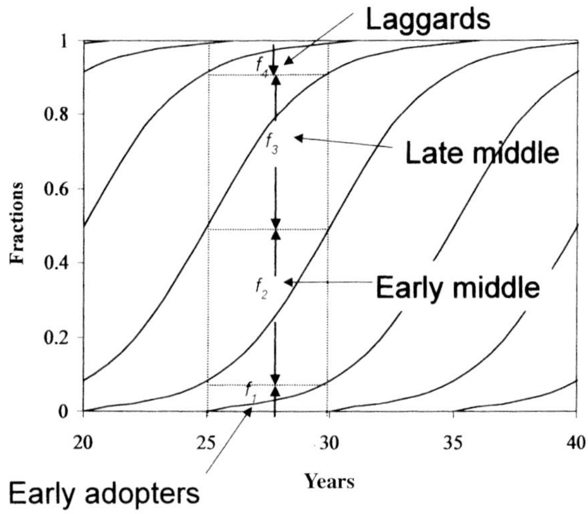

Figure 4. Logistic adaptation: Illustration of fraction of farmers (f) adapting to climate change during hypothetical period years 25–30. A small fraction of farmers (f_1) adapts immediately. Farmer fractions f_2, f_3, and f_4 adapt early middle, late middle and lagged, respectively. The fractions for f_1–f_4 can be derived from Equation (2), and the sum of f_1–f_4 equals 1.

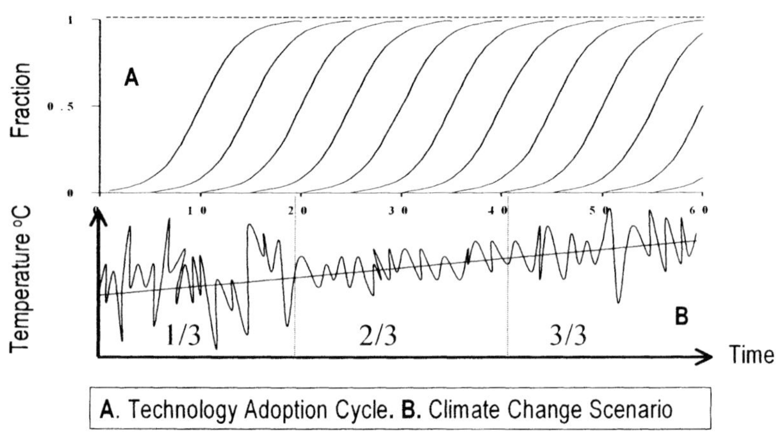

Figure 5. Technology substitution cycles (a) with hypothetical transient climate change (b).

which shows hypothetical 5-year waves of adaptation in the upper panel (A) versus a hypothetical transient climate change in the lower panel (B). In our case, we assume that a technology cycle spans 20 years, we get: $t_0 = 10$, and $\alpha = 0.48$ (see Appendix I for detailed procedures to calculate t_0 and α). Again, the logistic curve describing a full 20-year cycle is approximated by three 5-year computational time-steps. Adaptation cycles overlap, allowing for new adaptation to take place before previous cycles are complete (Figure 5). Such allows continual adjustment to adaptation in 5-year time steps but with varying degrees of efficiency among the subgroups f_1–f_4.

We use a matrix approach to combine model simulations at each of the 5-year time steps. Let **F** represent the matrix of fraction of farmers falling into the classes of clairvoyant (f_1), early middle (f_2), late middle (f_3) and lagged (f_4) adaptors, and $\mathbf{Y}_{yld}$, represents the matrix of yields from model simulations run with clairvoyant (y_{clv}), early middle (y_{ea}), late middle (y_{la}) and lagged ($_{lag}$) adaptations, respectively. A mean yield for logistic adaptations ($Y_{logistic}$) at a given 5-year step can be calculated from the following set of equations (Equations (2)–(5)). The elements f_i of **F** can be calculated from Equation (2). The elements of $\mathbf{Y}_{yld}$ (i.e., y_{clv}, y_{ea}, y_{la}, and y_{lag}) can be obtained from EPIC runs for each 5-year-step for adaptations discussed above.

$$fi = \int_{(i-1)*5}^{i*5} f'(t)dt \,, \tag{2}$$

where $f'(t)$ is the derivatives of Equation (1) and $i = 1, 2, 3, 4$.

$$\mathbf{F} = [f_1, f_2, f_3, f_4] \tag{3}$$

$$\mathbf{Y}_{yld} = [\mathbf{y}_{clv}, \mathbf{y}_{em}, \mathbf{y}_{la}, \mathbf{y}_{lag}] \tag{4}$$

$$\mathbf{Y}_{logistic} = \mathbf{F} * \mathbf{Y}^{-1}. \tag{5}$$

Four simulations are done for each grid box within a 20-year technology substitution cycle (assumed 20 years), meaning that the EPIC model was run in a 5-year interval, and a total of 12 runs were done for the entire period under study. As noted above, results are averaged over 20-year time periods and reported thusly.

3.3.4. *No Adaptation*

The no adaptation case is the industry standard for the so-called ‘dumb farmer’ scenario. The EPIC model is run with climate change using current technology and no agronomic adjustment, as if the farmer was oblivious to climate change. It is the least realistic of the scenarios, but we include it because it remains an assumption of a majority of the crop modeling studies in the literature and, thus, serves as a useful benchmark for comparison of our results (Gitay et al., 2001).

3.3.5. *Analytical Protocol*

Corn (maize) yields were simulated with the baseline climate and the ‘mock’ climate change at each grid box. The three different strategies of adaptation (clairvoyant, lagged, logistic) are compared with the no-adaptation case in order to evaluate the relative ameliorative effect of each strategy on simulated yields. First, we examine the mapped spatial distribution of climate change-induced yield changes under the three strategies of adaptation relative to yields simulated with the baseline climate. Second, we examine the mean yield changes under the three strategies aggregated to the levels of states in the region and then the entire study region. Third, we investigate the statistical significance of those yield changes using analysis of variance.

4. Results

4.1. SPATIAL DISTRIBUTION OF YIELD CHANGES

The spatial distribution of changes in corn yields (% of baseline yields at each third of climate change) for the four adaptation scenarios (no adaptation, clairvoyant, lagged, and logistic adaptation) generally tracked the distribution of the climate changes described above (Figures 6a–d, 7a–d, 8a–d). In the first 20 years (1/3 climate change), yield losses of up to –10% with no adaptation covered most of the northern half of the region while the southern half generally experienced yield gains of up to 10% (Figure 6d). As expected, clairvoyant adaptation was most effective in abating yield loss (Figure 6a), while lagged adaptation was least effective (Figure 6b). Clairvoyant adaptation almost totally eliminated yield losses in the first 20 years of climate change although it did not seem to stimulate yield gains beyond those of the no adaptation case. The pattern of yield change with logistic adaptation was indistinguishable from the lagged adaptation case and not as effective as clairvoyant adaptation (Figure 6c).

As the climate changes intensified in the second (2/3 climate change) and third (3/3 climate change) 20 years, the effectiveness of adaptation in terms of slowing the southward expansion of yield loss converged toward small differences (Figures 7a–d, 8a–d). In the no adaptation case, yield losses intensified in the north and spread southward during the 2/3 and 3/3 climate change. In the final 20 years of climate change, large clusters of grid boxes in the northern tier of the region experienced yield losses up to –40% regardless of adaptation scenario (Figures 8a–d). All three adaptation scenarios abated yield losses in the southernmost part of the region, with clairvoyant adaptation being most effective in promoting yield gains in that portion of the region (Figures 8a–d). The pattern of yield changes with logistic adaptation was virtually indistinguishable from lagged adaptation. The most important distinction in the distribution of yield changes between logistic and clairvoyant adaptation seems to be the larger degree to which the positive effects of climate change are assisted by clairvoyant adaptation than by logistic adaptation.

4.1.1. *Mean Yields for States*

Regional assessments of the impact of climate change often require state or provincial level information. Yields from each of the adaptation scenarios are presented as state-level percentages of baseline yields averaged over all three increments of climate change (Figures 9 and 10). The spatial variability seen in the mapped distributions is evident in the state-level means. Of the eight states in the study region, seven of them (Alabama, Georgia, Mississippi, Arkansas, North Carolina, South

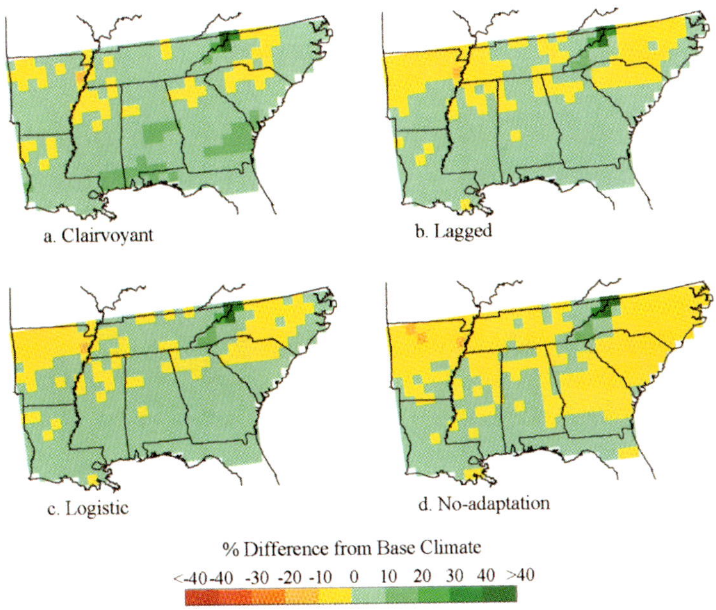

Figure 6. Response of corn yields (% of 1960–1995 baseline yields) to 1/3 of RegCM 2 × CO_2 climate change. (a) No adaptation. (b) Clairvoyant adaptation. (c) Lagged adaptation. (d) Logistic adaptation.

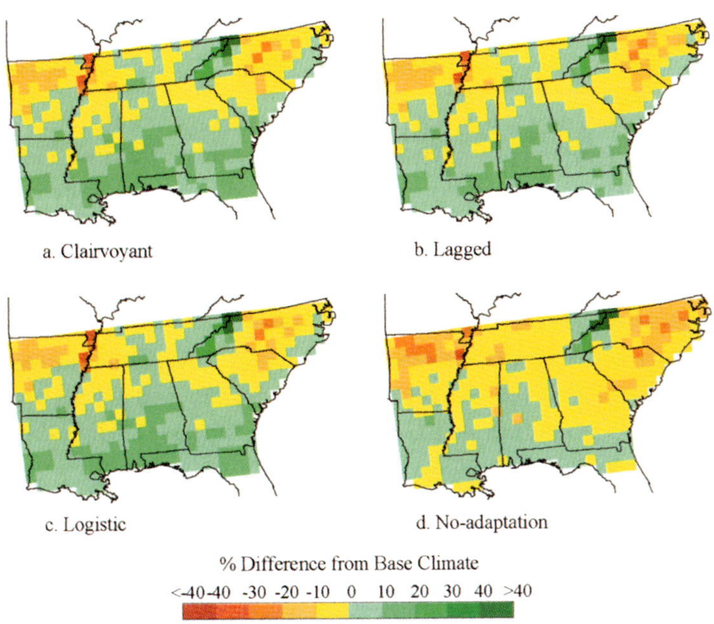

Figure 7. Response of corn yields (% of 1960–1995 baseline yields) to 2/3 of RegCM 2 × CO_2 climate change. (a) No adaptation. (b) Clairvoyant adaptation. (c) Lagged adaptation. (d) Logistic adaptation.

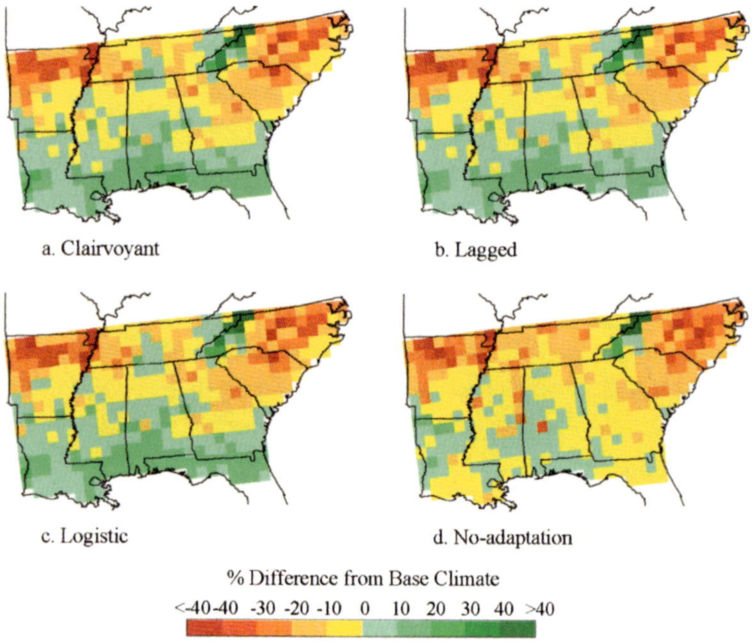

Figure 8. Response of corn yields (% of 1960–1995 baseline yields) to 3/3 of RegCM 2 × CO_2 climate change. (a) No adaptation. (b) Clairvoyant adaptation. (c) Lagged adaptation. (d) Logistic adaptation.

Carolina, and Tennessee) experienced yield losses relative to the baseline when no-adaptation is modeled (Figures 9 and 10). The exception is Louisiana. Each of the adaptation strategies (clairvoyant, logistic, and lagged) ameliorates at least a portion of the yield loss in all states. In four of the states (Alabama, Georgia, Mississippi, and Louisiana), all of the adaptation strategies show an increase in yield (Figures 9a–d). These states experienced yield gains of 4–7% with clairvoyant, 2–5% with logistic adaptation, and 1–4% with lagged adaptation. Although lagged adaptation was not as effective as clairvoyant adaptation, it clearly improved yields relative to the no-adaptation case.

However, none of the adaptation strategies could reverse the negative effects of the climate change (Figures 10a–d) in the northern part of the study region (Arkansas, North Carolina, South Carolina, and Tennessee). Here the yield loss ranges from –1% to –6% depending on the adaptation strategy.

4.2. MEAN CORN YIELD CHANGES ACROSS THE STUDY AREA

We posit that the effect of the technique used to model agronomic adaptation of corn yields to climate change on outcomes can be evaluated usefully at the scale of the entire southeastern U.S. We average the yields simulated with each adaptation strategy over all grid boxes to reach an overall mean for the study area. At each

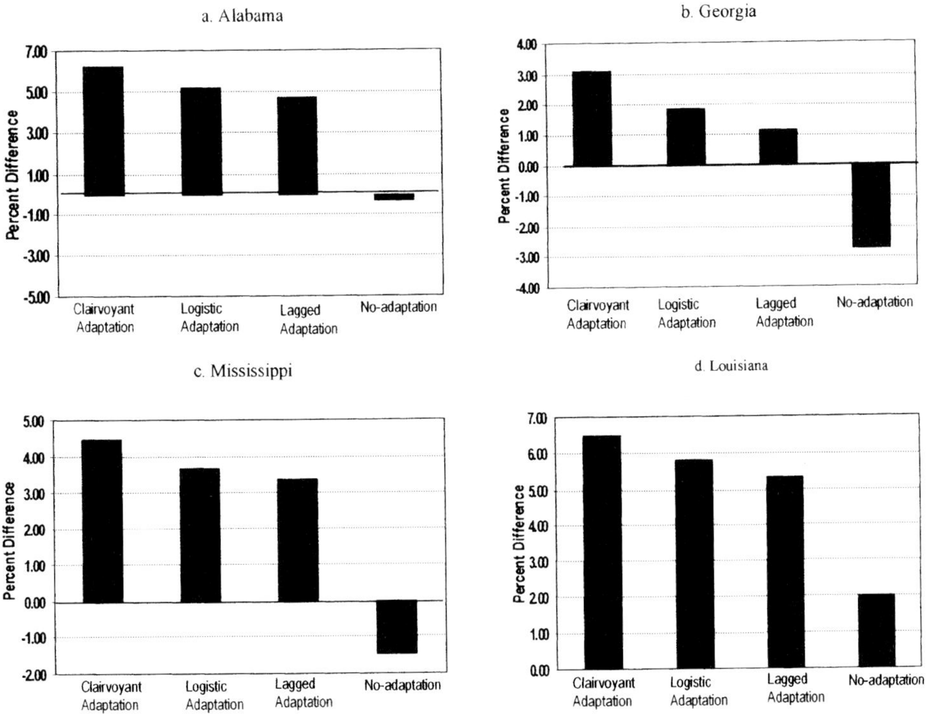

Figure 9. Mean response of corn yields (% of 1960–1995 baseline yields) over all three thirds of RegCM climate change in selected states. (a) Alabama. (b) Georgia. (c) Louisiana. (d) Mississippi.

third of climate change, we compare the mean yield for the no-adaptation strategy with the baseline (no climate change) yields; then we compare the yields simulated with each adaptation strategy to the no-adaptation case. Finally we compare yields computed with each adaptation strategy to yields with every other adaptation strategy. The point to this exercise is to determine the extent to which the assumptions and procedures for representing adaptation influence the conclusions reached.

4.2.1. *No Adaptation versus Baseline Yields*

Mean yield response to climate change with no-adaptation was compared with mean yield response to the 1960–1995 baseline climate in order to establish a benchmark for the adaptation simulations. In the first 20 years (first third) of climate change, mean corn yields with no adaptation averaged –0.2% less than mean baseline yields (Table I). By the second 20 years (second third) of climate change, yields with no adaptation were –2.5% less than baseline yields. By the third 20 years (third third), they dropped to –6.9%. This trend shows that, on balance, the effect of the climate change was to reduce yields in spite of substantial gains in yield in much of the southern portion of the study area (Figures 6–8).

Table II

Comparisons of grid-wise mean corn (maize) yield (mT/Ha and %) between no adaptation, lagged adaptation, logistic adaptation, and clairvoyant adaptation in response to 1/3, 2/3, and 3/3 climate change

Climate/adaptation scenarios	Mean yield (metric tons per hectare)	% Change in yield from no-adaptation scenario	% Change in yield from lagged adapt. scenario	% Change in yield from logistic adapt. scenario
1/3 RegCM climate change				
No adaptation	6.670			
Lagged adaptation	6.809	2.1		
Logistic adaptation	6.845	2.6 [b]	0.5	
Clairvoyant adaptation	7.002	5.0 [a]	2.8 [a]	2.3 [b]
2/3 RegCM climate change				
No adaptation	6.514			
Lagged adaptation	6.852	5.2 [a]		
Logistic adaptation	6.859	5.3 [a]	0.1	
Clairvoyant adaptation	6.920	6.2 [a]	1.0	0.9
3/3 RegCM climate change				
No adaptation	6.218			
Lagged adaptation	6.503	4.6 [a]		
Logistic adaptation	6.499	4.5 [a]	–0.1	
Clairvoyant adaptation	6.530	5.0 [a]	0.4	0.5

[a] Difference is statistically significant at $\alpha = 0.05$ level.
[b] Difference is statistically significant at $\alpha = 0.01$ but not at $\alpha = 0.05$ level.

our logistic adaptation methodology, that is the time when a substantial proportion of the farmers have yet to take defensive agronomic action with respect to climate change. In terms of lagged adaptation, that is also the time in which even a smaller proportion of farmers have made a move to adapt than with the logistic adaptation strategy.

The narrowing of the differences in yields among the adaptation strategies as the climate change strengthened indicates the relatively large benefit of even sub-optimal adaptation. The small (0.5%) difference between clairvoyant and logistic adaptation in the final third of climate change demonstrates the value of sub-optimal adaptation – this is underscored by both sets of yields being significantly different from the no adaptation case (Table II). That is, optimal (clairvoyant) and sub-optimal (logistic) adaptation substantially improve the no-adaptation yields, but are, themselves, not much different. Similar comparison of lagged and clairvoyant adaptation strengthens the case (Table II). The explanation for this finding lies in baseline agronomic practices (current planting dates, varietal growing season lengths) that are far from climatically optimal. The representative farm information used to prepare EPIC was derived of Cooperative Extension circulars and personal communication from Extension Educators. That information reflects the fact that farmers do their best to cope with climate, but balance several factors that influence agronomic practice in addition to climate considerations. The net effect is that farmers are usually slightly out of kilter with the climate. Moreover, a reasonable expectation of farmers' behavior with respect to climate change is their growing awareness of the change and an accompanying desire to become more optimally adapted to it over time than in the present.

In spite of the narrowing differences between adaptation strategies over the course of climate change, clairvoyance remained the superior strategy for maintaining the highest corn yields under progressive climate change. It establishes an upper bound of the envelope of adaptive potential and also provides a rationale for monitoring and detecting long-term climate change, and the vigilance of farmers and their supporting institutions. The implication of no adaptation as the lower bound of the envelope of response needs little elaboration. This envelope is important because there is no guarantee that these results are robust across an ensemble of climate change scenarios and in fact the methods of this research need to be tested against a range of climate change scenarios.

As noted above, the degree to which farmer behavior with respect to climate change can be expected to remain sub-optimal is an open question. The results of this analysis, however, suggest that delayed farmer response causes two problems for a region. First, inability to properly recognize and take appropriate steps to deal apace with the climate change builds in inefficiency proportional to the time elapsed from the point at which climate change commences to the point at which proper adaptation occurs. Second, the adaptive behavior of some farmers in a region will perpetually lag behind the current state of the climate change, putting those farmers in a quasi-permanent state of less-than-optimal efficiency. This could indicate a social welfare problem that may or may not have market solutions.

Over the long time scales of this analysis, simple substitutions of existing agronomic practices in response to climate change (of the kind tested in this research) sooner or later will lose effectiveness in dealing with climate change. This can be seen in the declining effectiveness of all three adaptation strategies in the final third of climate change as heat stresses overcome the value of planting longer season varieties (Table II). New fundamental adaptive knowledge and technology will be required to cope with unprecedented climate challenges. Cultivars bred or genetically modified to deal with changing climate conditions better than current ones can be expected from the private and public research communities. Many of the new innovations that will be needed to deal with climate change are currently over the horizon and will be radically new to future farmers. Although the process that generates those innovations is beyond the scope of this paper, it is reasonable to expect that the uptake of those innovations will more likely follow the logistic pathway than the suite of agronomic substitutions tested here for the simple reason that it will take time for the innovations to gain farmers' trust.

6. Conclusion

Adaptation to climate change has become an important consideration in climate change impact research. According to the recent IPCC Third Assessment Report (Gitay et al., 2001), nearly a hundred studies globally investigate the effects of adaptation on agricultural impacts of climate change and the number is growing. Our assessment is that few studies of agronomic adaptations evaluate the effect that the method by which the adaptations are applied has on outcomes. This is problematic.

We hypothesized that the method used to simulate agronomic adaptation has the potential to influence yields in response to climate change sufficient to alter outcomes and challenge the robustness of further (economic) modeling dependent on those yields. In particular, we posit that our use of the epidemiological (logistic) model of technological substitution as a method of allocating agronomic adaptation is more realistic than current methods *a priori* and should give significantly different yield outcomes from optimization-based methods. Our tests of these propositions with corn in the southeastern United States were inconclusive. Logistic adaptation did not produce significantly different yield outcomes from those of the more commonly used clairvoyant adaptation, except at the very outset of the warming. However, it is important to point out that this research used one climate change scenario. A complete test will require multiple climate change scenarios, crops and geographic locations.

We conclude that our formulation of logistic adaptation provides a disciplined and theoretically (and empirically) sound basis for modeling adaptations relying on technological innovation and substitution. Such is an advance, although the degree to which logistic adaptation may provide different policy-relevant information from common practice remains an open question for future research.

Acknowledgements

The authors wish to thank Linda Mearns for her constant vigilance in keeping the team harnessed and pushing forward, Robert Brown and Norm Rosenberg for providing EPIC data, Jon Voortman for help with data management, Cindy Hays for initial GIS programming, Larry McDaniel for providing climate data, and Mary Downton for her assistance with the statistical analysis. We also wish to acknowledge our many EMF-Snowmass Workshop colleagues, particularly Stephen H. Schneider, whose gentle cajoling inspired our thinking on adaptation.

Appendix I: Procedures Used to Derive the Constants t_0 and α in the Logistic Equation

1. t_0 is the inflection point at 0.5 of the fraction. It is right half of the time span. In our case, the total time period is 20 years, so $t_0 = 10$.
2. α is a constant regulating the growth rate or the steepness of the S-curve. The rate of growth is the time derivative of the S-curve, and generally follows a normal distribution over time. Based on the time derivative of the S-curve with a maximum value of M, the growth rage (GR) at time t_0 is given by (Debecker and Modis, 1994):

$$\mathrm{GR} = \frac{\mathrm{M}\alpha}{4}. \tag{6}$$

A general normal density function can be given as following equation (Harnett, 1982).

$$f(x) = \frac{1}{\sigma\sqrt{2\pi}} e^{-(1/2)[(x-\mu)/\sigma]^2} \tag{7}$$

and the following factor can be derived from formula (12) by substituting μ for x.

$$f(\mu) = \frac{0.399}{\sigma}. \tag{8}$$

Since μ in formula (12) is identical to t_0 in Equation (11), and $f(\mu)$ in Equation (13) is identical to GR in Equation (11), therefore:

$$\frac{\mathrm{M}\alpha}{4} = \frac{0.399}{\sigma}. \tag{9}$$

According to the characteristics of the normal distribution, $\mathrm{P}(\mu - 3\sigma < x < \mu + 3\sigma) = 0.9974$, we know that 6σ will cover almost all the time period (99.74%). So in our case, we set $6\sigma = 20$ (total time period 20 years for one technology cycle), and we derive $\sigma = 3.333$. By substituting 1 (maximum fraction), 3.333 for M and σ in Equation (14), respectively, we get:

$$\alpha = 0.48.$$

References

Brown, R. A. and Rosenberg, N. J.: 1999, 'Climate Change Impacts on the Potential Productivity of Corn and Winter Wheat in their Primary United States Growing Regions', *Clim. Change* **41**, 73–107.

CAST: 1992, *Preparing U.S. Agriculture for Global Climate Change*, Council for Agricultural Science and Technology, Ames, Iowa.

Debecher, A. and Modis, T.: 1994, 'Determination of the Uncertainties in S-Curve Logistic Fits', *Technological Forecasting and Social Change* **46**, 153–173.

Easterling, W. E.: 1996, 'Adapting North American Agriculture to Climate Change in Review', *Agric. For. Meteorol.* **80**, 1–53.

Easterling, W. E., Crosson, P. R., Rosenberg, N. J., McKenney, M., Katz, L. A., and Lemon, K.: 1993, 'Agricultural Impacts of and Responses to Climate Change in the Missouri-Iowa-Nebraska-Kansas (MINK) Region', *Clim. Change* **24**, 23–61.

Easterling, W. E., Mearns, L. O., Hays, C. J., and Marx, D.: 2001, 'Comparison of Agricultural Impacts of Climate Change Calculated from High and Low Resolution Climate Change Scenarios: Part II. Accounting for Adaptation and CO_2 Direct Effects', *Clim. Change* **51**, 173–197

Fisher, J. C. and Pry, R. H.: 1971, 'A Simple Substitution Model of Technological Change', *Technological Forecasting and Social Change* **3**, 75–88.

Frederick, K. D and Schwarz. G.: 1999, 'Socioeconomic Impacts of Climate Change on U.S. Water Supplies', *Journal of the American Water Resources Association* **35**, 1563–1583.

Geroski, P. A.: 2000, 'Models of Technology Diffusion', *Research Policy* **29**, 603–625.

Gitay, H., Brown, S., Easterling, W. and Jallow, B. (with Antle, J., Apps, M., Beamish, R., Chapin, T., Cramer, W., Frangi, J., Laine, J., Lin Erda, L., Magnuson, J., Noble, I., Price, J., Prowse, T., Sirotenko, O., Root, T., Schulze, E., Sohngen, B., and Soussana, J.): 2001, 'Ecosystems and their Goods and Services', in McCarthy et al. (eds.), *Climate Change 2001 – Impacts, Adaptations and Mitigation of Climate Change: Scientific-Technical Analyses*, Contribution of Working Group II to the Third Assessment Report of the Intergovernmental Panel on Climate Change, United Nations Environment Programme-World Meteorological Organization, Cambridge University Press, pp. 235–342.

Grübler, A.: 1998, *Technology and Global Change*, Cambridge University Press, Cambridge, pp. 50–75.

Harnett, D. L.: 1982, *Statistical Methods*, 3rd edn., Addison-Wesley Publishing Company, Inc.

Kaiser, H. M., Riha, S. J., Wilks, D. S., Rossier, D. G., and Sampath, R.: 1993, 'A Farm Level Analysis of Economic and Agronomic Impacts of Gradual Warming', *Amer. J. Agric. Econ.* **75**, 387–398.

Kiniry, J. R, Jones, C. A., O'Toole, J. C., Blanchet, R., Cabeiguenne, M., and Spanel, D. A.: 1989, 'Radiation Use Efficiency in Biomass Accumulation Prior to Grain Filling in Five Grain-Crop Species', *Field Crops Reasearch* **20**, 51–64.

Mansfield, E.: 1968, *Industrial Research and Technological Innovation*, W. W. Norton, New York.

Mearns, L. O., Giorgi, F., McDaniel, L., and Shields, C.: 2003, 'Climate Scenarios for the Southeastern U.S. Based on GCM and Regional Model Simulations', *Clim. Change* **60**, 7–35.

Mearns, L. O., Easterling, W., Hays, C., and Marx, D.: 2001, 'Comparison of Agricultural Impacts of Climate Change Calculated from High and Low Resolution Climate Change Scenarios: Part I. The Uncertainty Due to Spatial Scale', *Clim. Change* **51**, 131–172.

Parry, M. L., Carter, T. R., and Konijn, N. T. (eds.): 1988, *The Impact of Climatic Variations on Agriculture*, Kluwer, Dordrecht.

Reilly, J. M.: 1995, 'Climate Changes and Global Agriculture: Recent Findings and Issues', *Amer. J. Agric. Econ.* **77**, 727–733.

Rosenberg, N. J.: 1982, 'The Increasing CO_2 Concentration in the Atmosphere and its Implication on Agricultural Productivity: Part II. Effects through CO_2-Induced Climatic Change', *Clim. Change* **4**, 239–254.

Rosenzweig, C. and Parry, M.: 1994, 'Potential Impact of Climate Change on World Food Supply', *Nature* **367**, 133–138.

Ruttan, V. W.: 1996, 'What Happened to Technology Adoption-Diffusion Research?', *Sociologia Ruralis* **36**, 51–73.

Schneider, S. H., Easterling, W. E., and Mearns. L. O.: 2000, 'Adaptation: Sensitivity to Natural Variability, Agent Assumptions, and Dynamic Climate Changes', *Clim. Change* **45** (1), 203–221.

Smit, B., McNabb, D., and Smithers, J.: 1996, 'Agricultural Adaptation to Climate Variation', *Clim. Change* **33**, 7–29.

Stockle, C. O., Williams, J. R., Rosenberg, N. J., and Jones, C. A.: 1992, 'A Method for Estimating Direct and Climatic Effects of Rising Atmospheric Carbon Dioxide on Growth Yield of Crops: Part I – Modification of the EPIC Model for Climate Change Analysis', *Agricultural Systems* **38**, 225–228.

Williams, J. R., Jones, C. A., and Dyke, P. T.: 1984, 'A Modeling Approach to Determining the Relationship between Erosion and Soil Productivity', *Trans. Am. Soc. Agric. Eng.* **27**, 129–144.

Williams, J. R., Jones, C. A., Kiniry, J. R., and Spaniel, D. A.: 1989, 'The EPIC Crop Growth Model', *Trans. Am. Soc. Agric. Eng.* **32**, 497–511.

Yohe, G. W. and Schlesinger, M. E.: 1998, 'Sea Level Change: The Expected Economic Cost of Protection or Abandonment in the United States', *Clim. Change* **38**, 447–472.

(Received 29 May 2002; in revised form 21 April 2003)

SOUTHEASTERN U.S. VEGETATION RESPONSE TO ENSO EVENTS (1989–1999) *

ALBERT J. PETERS [1], LEI JI [1] and ELIZABETH WALTER-SHEA [2]

[1] *Center for Advanced Land Management Information Technologies, University of Nebraska-Lincoln, 113 Nebraska Hall, Lincoln, NE 68588-0517, U.S.A.*
E-mail: apeters@calmit.unl.edu

[2] *School of Natural Resource Sciences, University of Nebraska-Lincoln, 303 Biochemistry Hall, Lincoln, NE 68583-0758, U.S.A.*

Abstract. El Niño/Southern Oscillation (ENSO) is considered one of the most powerful forces driving anomalous global weather patterns. Large-scale seasonal precipitation and temperature changes influenced by ENSO have been examined in many areas of the world. The southeastern United States is one of the regions affected by ENSO events. In this study, remote sensing detection of vegetation response to ENSO phases is demonstrated with one-kilometer biweekly Normalized Difference Vegetation Index (NDVI) data (1989–1999) derived from the Advanced Very High Resolution Radiometer (AVHRR). The impacts of three ENSO phases, cold, warm and neutral, on vegetation were analyzed with a focus on two vegetation cover types, two seasons and two geographic regions within the southeastern U.S. Significant ENSO effects on vegetation were found in cropland and forest vegetation cover types based on image and statistical analysis of the NDVI data. The results indicate that vegetation condition was optimal during the ENSO neutral phase for both agricultural and natural vegetation.

1. Introduction

El Niño-Southern Oscillation (ENSO) is a coupled ocean-atmosphere phenomena in the tropical Pacific that is, after the seasons, the most powerful force driving global weather changes. Large-scale precipitation and temperature patterns influenced by ENSO have been examined in many areas of the world (Ropelewski and Halpert, 1986; Kiladis and Dias, 1989; Barnett et al., 1994; Adams et al., 1995; Barsugli et al., 1999; McCabe and Dettinger, 1999; Rajagopalan et al., 2000). The Southeast United States is no exception. For example, during warm events, winter precipitation is anomalously high, but temperatures are low due to increased cloud cover, while summers tend to be dry along the southeast coast (Ropelewski and Halpert, 1986; Adams et al., 1995). Impacts of ENSO on agriculture and crop production in the Southeast U.S. have been reported (Adams et al., 1995; Hansen et al., 1998).

Satellite imagery provides high-spatial and -temporal resolution coverage, which can be used to efficiently evaluate global and regional vegetation condition.

* Published as Paper No. 13698, Journal Series, Nebraska Agricultural Research Division.

Climatic Change **60:** 175–188, 2003.

At present the Advanced Very High Resolution Radiometer (AVHRR) on board the National Oceanic and Atmospheric Administration's (NOAA's) polar-orbiting satellites is the instrument of choice for collecting 1-kilometer resolution imagery worldwide because there is more than 12 years of historical data (Batista et al., 1997). The Normalized Difference Vegetation Index (NDVI), a measure typically derived from AVHRR data, is calculated as:

$$\mathrm{NDVI} = (\rho_{\mathrm{NIR}} - \rho_{\mathrm{vis}})/\rho_{\mathrm{NIR}} + \rho_{\mathrm{vis}}),$$

where ρ_{NIR} is the reflectance in the near-infrared waveband and ρ_{vis} is the reflectance in the visible red waveband of the satellite radiometer (Justice et al., 1985). This ratio provides an indicator of vegetation condition, or vigor, such that the higher the NDVI value the greater the level of photosynthetic activity in the cover type (Sellers, 1985; Tucker et al., 1991: Bannari et al., 1995; Tucker and Sellers, 1996). Multitemporal NDVI derived from AVHRR data are useful for monitoring vegetation dynamics on regional and continental scales (Goward et al., 1985; Tucker et al., 1991, 1996; Eidenshink and Haas, 1992; Benedetti and Rossini, 1993; Unganai and Kogan, 1998; Moulin et al., 1998). Therefore, NDVI derived from AVHRR can be applied effectively to detect variations in vegetation condition induced by ENSO at global and continental scales.

Recent research has investigated the relationship between AVHRR-NDVI and ENSO indices in various regions of the world. For example, Eastman and Fulk (1993), and Anyamba and Eastman (1996) employed principal component analysis techniques for time series AVHRR-NDVI in Africa. They found that interannual NDVI variation was strongly related to three ENSO indices: atmospheric pressure patterns, Pacific sea-surface temperature and anomalous outgoing long-wave radiation. Significant reductions in NDVI values were associated with ENSO events in the Amazonian region of South America (Batista et al., 1997). Gutman et al. (2000) used AVHRR data to determine the signatures of ENSO impacts on the surface-atmosphere system over the Indonesian archipelago. Kogan (2000) examined the association between sea surface temperatures (SST) anomalies in the tropical pacific and a vegetation health index derived from AVHRR-NDVI for the entire world. Liu and Engrón Juárez (2001) constructed linear regression models for AVHRR-NDVI and ENSO indices and applied the models to predict ENSO-related drought onset in Brazil. Li and Kafatos (2000) found the interannual variation of vegetation estimated from AVHRR-NDVI is strongly related to ENSO indices in the United States. Mennis (2001) found a clearly recognizable relationship between ENSO events and interannual fluctuation in NDVI in the states of North and South Carolina, and Georgia in the southeastern U.S. Warm phase events were associated with a decline in vegetation vigor while the cold phases were associated with an increase.

So far, studies of ENSO impact in the southeastern U.S. have focused primarily on the response of climate (Ropelewski and Halpert, 1986; Rajagopalan et al., 2000; Goodman et al., 2000; Katz et al., 2003) and agricultural yield (Adams et al., 1995, 1999; Hansen et al., 1998). Investigations into the response of vegetation as viewed with satellite data have been reported recently (Kogan, 2000; Li and Kafatos, 2000; Mennis, 2001). Although Kogan (2000) and Li and Kafatos (2000) found ENSO impacts on land ecosystems in the southeastern U.S., their interests were focused on NDVI variation at global and continental scales. Mennis (2001) looked at the relationship between SST and NDVI, noting a negative correlation between SST anomalies and NDVI, in the states of North and South Carolina, and Georgia, but regional variability of NDVI in response to ENSO was not demonstrated. The objective of our study was to investigate changes in the AVHRR signal as an indicator of changes in vegetation condition, induced by ENSO in the southeastern U.S. as a whole and variation within the region as well, with fine spatial resolution imagery. Two cover types, two seasons and two geographic areas were compared and contrasted in order to explore ENSO impacts on vegetation.

2. Material and Methods

2.1. DATA

2.1.1. *Satellite Sensor Data*

Biweekly (14-day) Maximum Value Composite NDVI (1-km resolution) data (1989–1999) from the Earth Resources Observation System (EROS) Data Center, United States Geological Survey (USGS) were used in this study. The maximum value composite process retains the highest NDVI values during each 14-day period, producing images that are spatially continuous and relatively cloud-free, with temporal resolution sufficient for evaluating vegetation dynamics (Holben, 1986; Eidenshink, 1992). Data from September–December 1994 was not used due to failure of the AVHRR sensor.

2.1.2. *Seasonal Land Cover*

The Conterminous U.S. Land Cover Characteristics Data set (Loveland et al., 1991; Eidenshink, 1992), developed at EROS and the University of Nebraska-Lincoln (U.S.A.) was used to characterize land cover based on seasonal vegetation phenology. This land cover image was developed with multi-temporal AVHRR time-series from 1990. Post-classification refinement was accomplished using ancillary earth science data sets, including topography, climate, soils, and ecoregions. Land cover classes are defined at 1 kilometer. For our purposes, the original 159 classes were regrouped into seven cover types (Figure 1), and five were used for this analysis: cropland, mixed cropland/woodland and deciduous, coniferous and mixed forests. From these five classes, two broad classes were defined: cropland (cropland and mixed cropland/woodland classes) and forest (deciduous, coniferous and mixed forest classes). Pixels classified as urban and built-up areas, wetlands, water bodies and 'other' were removed from the analysis to reduce the impact of these cover types on those of interest.

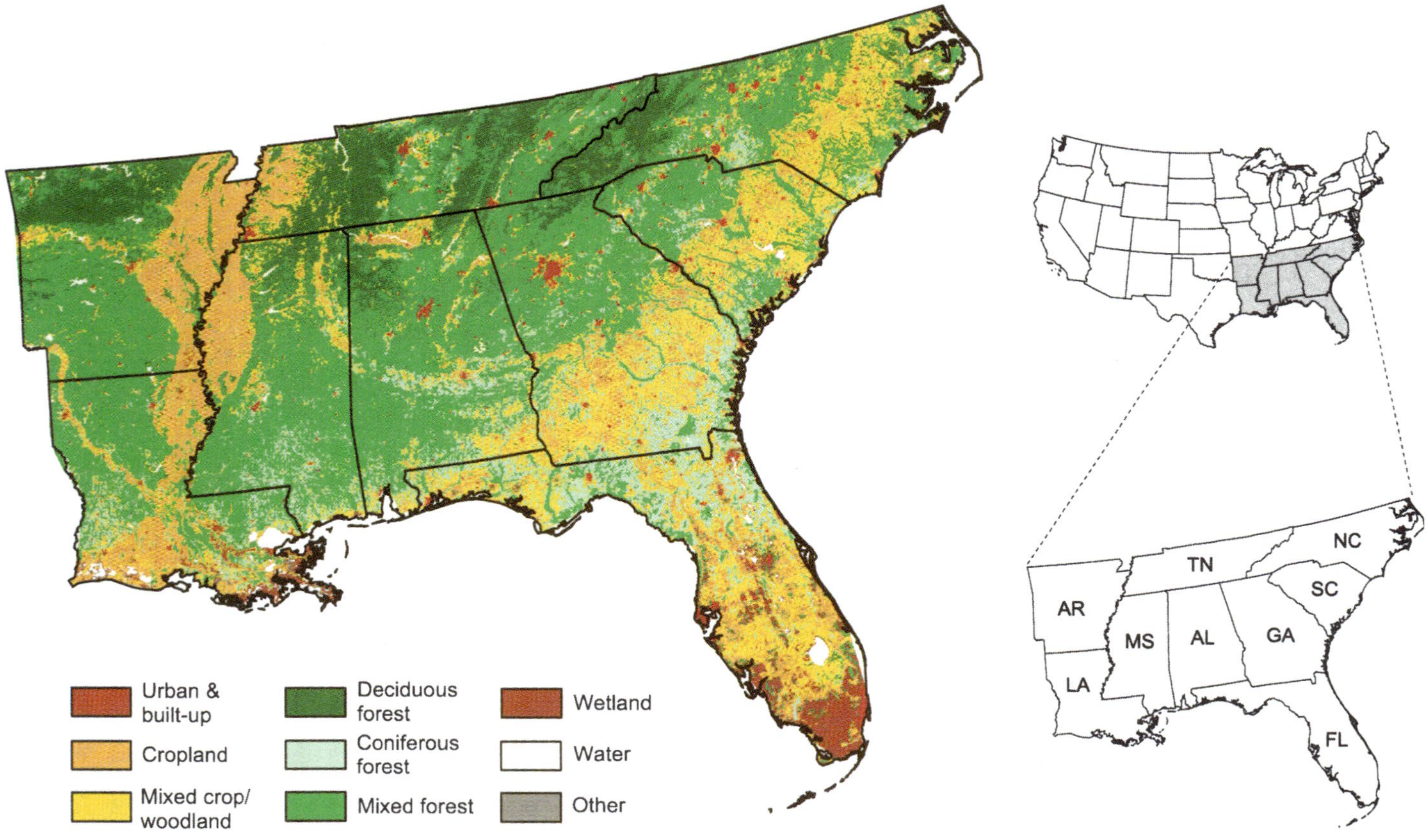

Figure 1. Land cover in the southeastern U.S. compiled from the Conterminous U.S. Land Cover Characteristics database (Loveland et al., 1993). Percent of pixels in each cover type: Urban and built-up (1.4%), cropland (10.9%), mixed cropland/woodland (20.9%), deciduous forest (11.3%), coniferous forest (10.1%), mixed forest (40.6%), wetland (1.8%), water (1.9%), and other (0.2%).

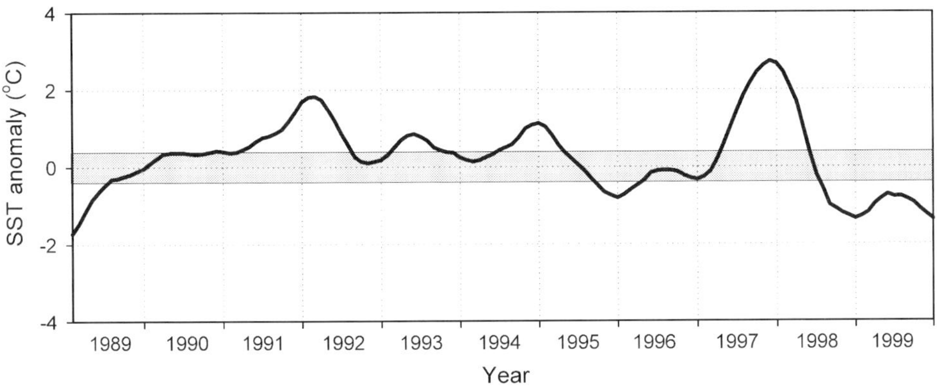

Figure 2. Time Series of SST (Niño 3.4) anomalies (1989–1999). The gray bar represents ENSO phase thresholds at ±0.4 °C.

2.1.3. *Sea Surface Temperature*

Sea Surface Temperatures (SSTs) in the equatorial Pacific Ocean were used to define ENSO events. By definition, El Niño occurs if 5-month running means of SST anomalies in the Niño 3.4 region (120° W–170° W, 5° S–5° N) exceed 0.4 °C for 6 months or more (Trenberth, 1997). SST anomalies were calculated by subtracting the monthly SST mean from the monthly SST time series. The monthly SST mean was derived from a base period of 1950–1979. Trenberth and Hoar (1996) stated that this base period was representative of sea-surface temperature conditions during this century, whereas the period after 1979 had been biased by warm events. SST data obtained from the Climate Prediction Center, NOAA/National Weather Service (http://www.cpc.noaa.gov/data/indices/index.html) were used to calculate the 1989–1999 monthly SSTs (Figure 2). Three ENSO phases were defined based upon the SST anomalies: warm and cold phases are defined as temperature anomalies above and below ±0.4 °C respectively, while the neutral phase is defined when temperatures are within the range of ±0.4 °C (Figure 3).

2.2. METHODS

2.2.1. *Standardized Departure of NDVI*

Monthly NDVI images were created using a linear interpolation technique in order to match the SST time series. Standardized Departure of NDVI was used to permit viewing the spatial patterns of vegetation change during ENSO phases. This is defined as the difference between the mean NDVI during an ENSO phase and the mean NDVI of the study period (1989–1999), divided by the standard deviation of the time series. Standardized departure images were calculated as the average departure during the growing season (May–October) and winter (December–February) for each ENSO phase (1989–1999) so that trends for each ENSO phase could be observed. Standardized departure could thus be negative (indicating a decrease in vegetation vigor), positive (indicating an increase) or close

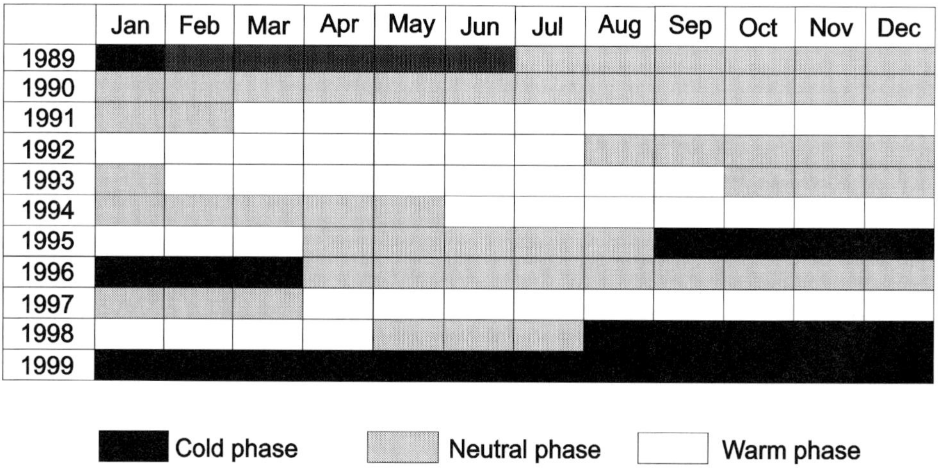

Figure 3. Cold, neutral and warm ENSO phases as defined by SSTs in the Niño 3.4 Region.

to zero (little or no change). Trends in NDVI departure were compared to climate observations from the *Weekly Weather and Crop Bulletin* (NOAA).

2.2.2. *Regional Significance of ENSO Phases in Cropland and Forests*

To geographically isolate ENSO impacts in the SE U.S., two subregions were selected for analysis, the Gulf Coast and Interior; an 'other' class included remaining areas (Figure 4). Boundaries of these areas are defined by Climate Divisions. Time-series of cropland- and forest-NDVI means were retrieved for each biweekly period from 1989–1999. The mean and standard deviation of NDVI by ENSO phase and subregion were calculated for each Climate Division. Analysis of variance (ANOVA) was used to test the significance of ENSO phase effects on regional NDVI. In a two-way ANOVA model (Kuehl, 2000), ENSO phase and regional effects were considered as factors affecting vegetation condition. The linear ANOVA model for the two-way factorial design is given by:

$$\mathrm{NDVI}_{ijk} = \mu + \mathrm{ENSO}_i + \mathrm{subregion}_j + (\mathrm{ENSO} \times \mathrm{subregion})_{ij} + \varepsilon_{ijk}\,,$$

where μ is the overall mean NDVI for all ENSO phases for the entire southeastern U.S., ENSO_i is the effect of the ith ENSO phase (cold, warm, neutral), $\mathrm{subregion}_j$ is the effect of the jth subregion (gulf coast, interior, other), $(\mathrm{ENSO} \times \mathrm{subregion})_{ij}$ is an interaction term between ENSO and subregion, and ε_{ijk} is random error associated with the ith ENSO and jth subregion.

Spatial autocorrelation of NDVI likely exists between climate divisions since there are obvious spatial gradients and heterogeneity. Spatial autocorrelation violates the basic assumption of independent observations in a conventional ANOVA and can result in erroneous conclusions about the effect of the factors (Legendre, 1993; Griffith, 1996). Thus, an adjustment for spatial autocorrelation was applied by integrating this variability into the ANOVA model with the mixed model tech-

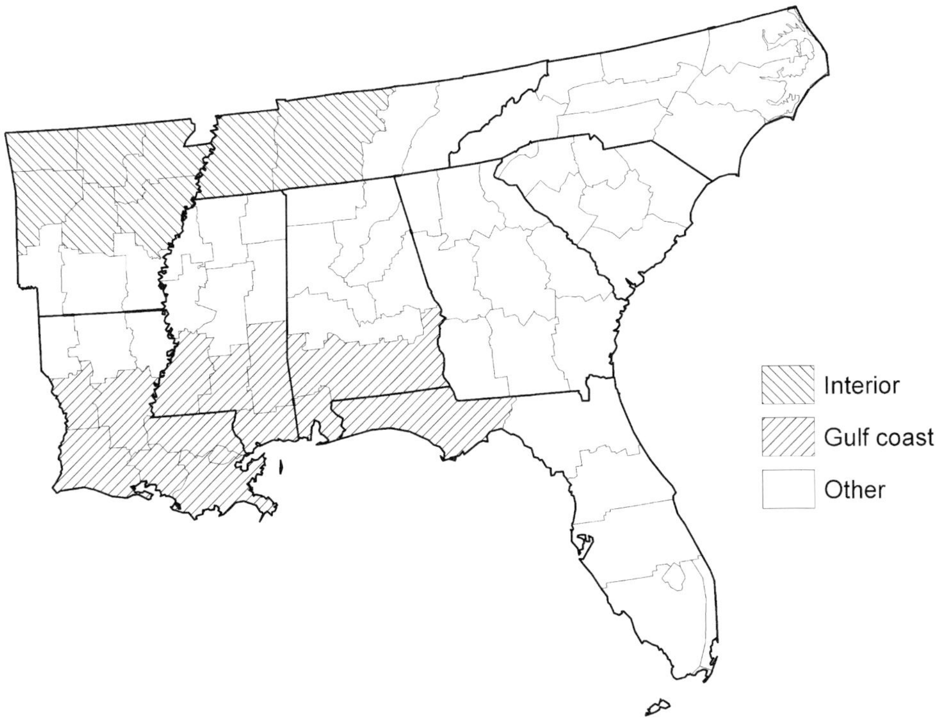

Figure 4. Interior, Gulf Coastal and other subregions of the Southeastern U.S. with boundaries defined by the Climate Divisions.

nique (Littell et al., 1996). The least square means were then adjusted and new NDVI values were compared by region, season and ENSO phase. The ANOVA was performed for each combination of the two cover types and two seasons, i.e., cropland/growing season, forest/growing season, cropland/winter and forest winter. The ENSO, subregion and their interaction effects were tested with the ANOVA.

3. Results and Discussion

3.1. OVERALL VIEW OF NDVI RESPONSE TO ENSO PHASES

A comparison of SST in the Niño 3.4 region (Figures 2 and 3) and NDVI standardized departure images in the southeastern U.S. (Figure 5) depict the trends and spatial variation in vegetation response to ENSO phases. The SST warm phases tend to be characterized by summer droughts, resulting in lower than average NDVI (Figure 5C). The years 1991–94 represent an SST warm phase interrupted by brief periods of SST neutral conditions, and were characterized by reduced crop yields and dry pastures during the summer in most areas and extreme winter moisture and increased NDVI along the Gulf Coast and Florida. In fact, 1993 was the driest and the second hottest summer of the 20th century in the Southeast. The year 1997 was characterized by drought aggravated by spells of heat that hurt crops in parts of the Southeast. The finding of reduced vegetation condition during the warm phase is corroborated by the work of Mennis (2001) who concluded that the vegetation vigor of all land cover classes declined during the warm phase.

Cold SST anomalies tended to be characterized by pockets of drought as indicated by poor vegetation condition (Figure 5A). Lingering dryness in 1989 in the aftermath of the severe 1988 drought was evident; however, timely summer rains in 1989 returned corn and soybean crops to their near long-term yield trend. In 1995 a marked change in circulation patterns resulting from the decline of the long warm phase, brought severe heat waves through September. Diminishing wet and stormy conditions in the early part of 1998 coincided with declining sea surface temperatures and resulted in improved growing conditions. Drought occurred in the summer but was eased somewhat by scattered thunderstorms. However, winter dryness continued into 1999 through the summer, causing considerable losses for agriculture as growing conditions deteriorated during the cold SST conditions that had not seen since 1988/1989. The cold phase growing season exhibited a complex spatial pattern with below average vegetation conditions throughout Florida, along the Mississippi River floodplain and the North Carolina coast while above average conditions were observed along the Gulf Coast centered in Mississippi and Louisiana (Figure 5A). The Mississippi floodplain is one of the most densely cropped sections of the study area (Figure 1), where rice, soybeans, cotton and winter wheat are grown. The negative impact of the cold phase vs. warm phase weather on cropland in this area can be seen in Figures 5A,C.

Average to above average growing-season NDVI occurred during the SST neutral phases (Figure 5B). These conditions occurred in late 1989 and early 1990 and in brief periods of 1992, 1994, 1996 and 1998 (Figures 2 and 3). Late 1989 was characterized by wet conditions which broke the drought and resulted in improved growing conditions, with corn and soybean yields back to near normal. Higher crop yields in the Mississippi valley in neutral as compared to cold phase years have been confirmed (http://www.usda.gov/nass/graphics). Of the three SST anomalies, the neutral phase was optimum for vegetation during the growing season. This differs from previous results found in correlation analyses of SST and NDVI relationships (Kogan, 2000; Mennis, 2001; Liu and Juárez, 2001), due primarily to the fact that these relationships are not linear. In other words, both high and low SST anomalies induce less favorable climate for vegetation with the warm phase having the greater negative effect.

Winter growing conditions also showed a considerable amount of variation between ENSO phases. During neutral SST conditions (Figure 5E), NDVI was fairly close to average throughout the region with the exception of the Gulf Coast and Florida. Cold phase NDVI (Figure 5D) was generally above average which is consistent with warmer and drier winter weather (Ropelewski and Halpert, 1986;

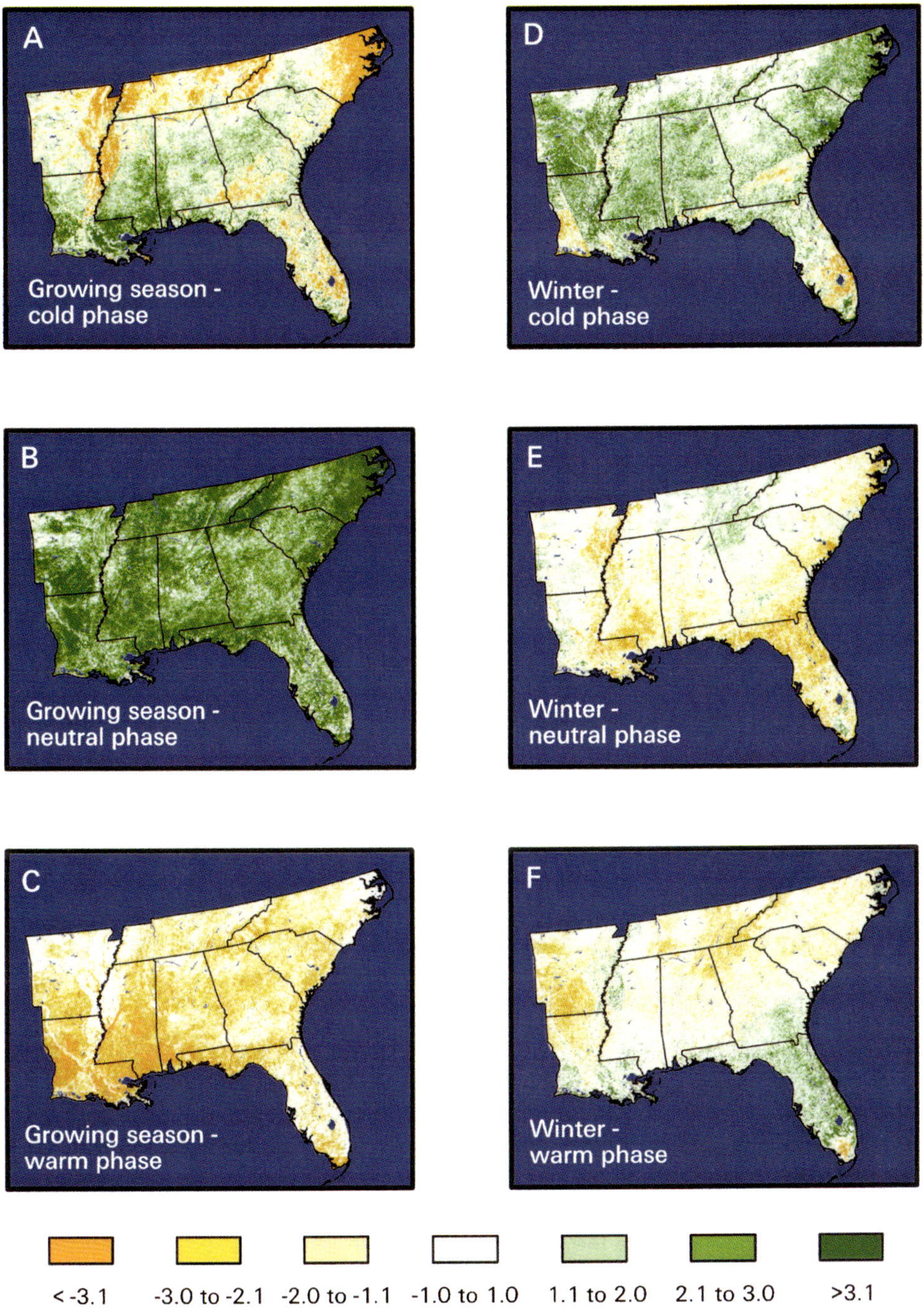

Figure 5. Standardized Departure of NDVI by ENSO phases: A: Growing season – cold; B: Growing season – neutral; C: Growing season – warm; D: Winter – cold; E: Winter – neutral; F: Winter – warm.

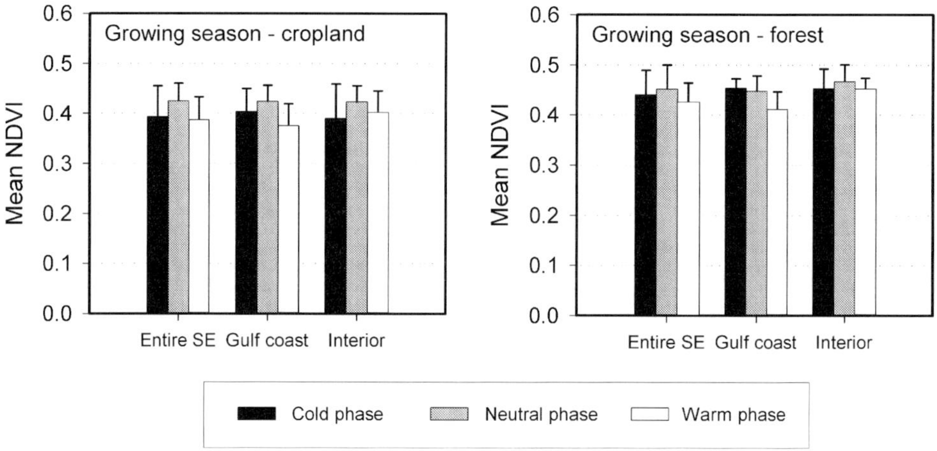

Figure 6. Mean NDVI for the three ENSO phases. Comparisons are made for cropland vs. forest and gulf coast vs. interior regions during the growing season from May thru October. The horizontal bars indicate one standard deviation.

Kiladis and Diaz, 1989). Warm-phase winter vegetation was average to below average except along the Gulf Coast and most of Florida (Figure 5F), which is consistent with more sun and drier weather in this region (Ropelewski and Halpert, 1986; Kiladis and Diaz, 1989). Warm phase winters are characterized by wet and mild weather, resulting in lower than average NDVI, except along the Gulf coast (Figures 5C,F).

3.2. IMPACT OF CONTINENTALITY ON ENSO PHASES

Another aspect of spatial variation in vegetation response can be demonstrated when the study area is divided into three subregions for comparison (Figure 4). The mean and standard deviation of NDVI for each combination of these subregions and ENSO phases provided the basis for the analysis (Figure 6). The effects of ENSO and subregion on the NDVI can be visualized by comparing the mean NDVI values between the combinations. Based on these comparisons, growing season vegetation is optimal in entire Southeastern U.S. and in all subregions during the neutral phase.

The ANOVA provides the statistical test of the effects of ENSO and subregion on NDVI, as summarized in Figure 7. The p-value in the figure is the probability of rejecting the null hypothesis that there is no difference between the combinations. A p-value less than 0.05 indicates that there is a significant difference between subregions in terms of NDVI response. The adjusted least-squares-mean (Littell et al., 1996) of NDVI is the mean value predicted with the ANOVA model for each ENSO phase and/or subregion, adjusted for spatial autocorrelation between climate divisions.

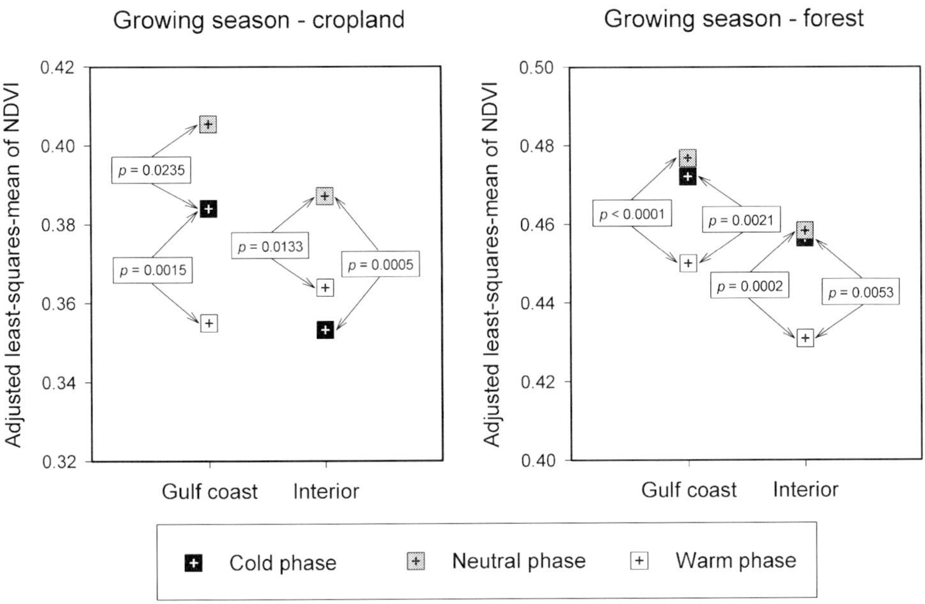

Figure 7. Comparison of ENSO phase and subregion effects on NDVI derived from the ANOVA. The boxes with p-values indicate the difference between two groups at $p < 0.05$, suggesting that the difference between the two groups is significant.

The ANOVA results indicate that ENSO effects were significant in both the Gulf and Interior. Vegetation condition was the best during neutral SST anomalies regardless of location or type. This is attributed to the more favorable growing conditions. Along the coast, neutral and cold phase vegetation (crop and forest) was significantly better than during the warm phase, which can be attributed to the summer dryness during these periods. Vegetation response was similar both in the Interior and along the coast. That is, crops were better during neutral, relative to cold phase conditions. Forest conditions were also better during neutral and cold phases than during the warm phase.

The ANOVA results show that there is no significant interaction between ENSO and subregion for cropland ($p = 0.0895$) and for forest (0.7347). In other words, there is no significant difference between Gulf coast and interior areas in terms of the ENSO impact. This suggests that variation in SSTs affects vegetation vigor in the entire Southeast, not just the Gulf coast area.

4. Conclusions

A vegetation response to ENSO phases was detected using AVHRR NDVI and SST information. NDVI varied between ENSO phases, type of vegetation and location (Interior vs. Gulf Coast). Based on the analyses the following conclusions can be made:

- The neutral ENSO phase resulted in above average vegetation response, suggesting that the neutral phase provides optimal climate conditions for vegetation in the Southeastern U.S.
- The warm ENSO phase resulted in the lowest NDVI, indicating a negative effect on vegetation attributed to the cloudy-wet winters and extreme summer weather events.
- Variations in vegetation response do not strongly depend on location. ENSO has the similar influence on vegetation conditions along the Gulf Coast and in the Interior.

This study provides the most detailed investigation to date of ENSO impacts on vegetation detected with satellite sensor data in the southeastern U.S. The results of the study should promote understanding of ENSO teleconnections between the atmosphere and vegetation in the southeast U.S., and provide information to help predict ENSO effects on vegetation.

Acknowledgements

This research was supported by a NASA MTPE Grant No. OA99073, W-19, 080. We thank the two anonymous reviewers for their helpful comments.

References

Adams, R. M., Chen, C.-C., McCarl, B. A., and Weiher, R. F.: 1999, 'The Economic Consequences of ENSO Events for Agriculture', *Clim. Res.* **13**, 165–172.

Adams, R. M., Kelly, K. J., McCarl, B. A., Legler, D. M., Obrian, J. Solow, A., and Weiher, R.: 1995, 'Value of Improved Long-Range Weather Information', *Contemporary Economic Policy* **13**, 10–19.

Anyamba, A. and Eastman, J. R.: 1996, 'Interannual Variability of NDVI over Africa and its Relation to EL Niño/Souththern Oscillation', *Int. J. Remote Sensing* **17**, 2533–2548.

Bannari, A., Morin, D., and Bonn, F.: 1995, 'A Review of Vegetation Indices', *Rem. Sens. Rev.* **13**, 95–120.

Barnett, T. P., Bengtsson, L., Arpe, K., Flugel, M., Graham, N., Latif, M., Ritchie, J., Roeckner, E., Schlese, U., Schulzweida, U., and Tyree, M.: 1994, 'Forecasting Global ENSO-Related Climate Anomalies', *Tellus* **46A**, 381–497.

Barsugli, J. J., Whitaker, J. S., Loughe, A. F., Sardeshmukh, P. D., and Toth, Z.: 1999, 'The Effect of the 1997/98 El Niño on Individual Large-Scale Weather Events', *Bull. Amer. Meteorol. Soc.* **80**, 1399–1408.

Batista, G. T., Shimabukuro, Y. E., and Lawrence, W. T.: 1997, 'The Long-Term Monitoring of Vegetation Cover in the Amazonian Region of Northern Brazil Using NOAA-AVHRR Data', *Int. J. Remote Sensing* **18**, 3195–3210.

Benedetti, R. and Rossini, P.: 1993, 'On the Use of NDVI Profiles as a Tool for Agricultural Statistics: The Case Study of Wheat Yield Estimate and Forecast in Emilia Romagna', *Rem. Sens. Env.* **45**, 311–326.

Eidenshink, J. C. and Hass, H. R.: 1992, 'Analyzing Vegetation Dynamics of Land Systems with Satellite Data', *Geocarto International* **7**, 53–61.

Eidenshink, J. C.: 1992, 'The 1990 Conterminous U.S. AVHRR Data Set', *Photogrammetric Engineer. Remote Sens.* **58**, 809–813.

Eastman J. R. and Fulk, M.: 1993, 'Long Sequence Time Series Evaluation Using Standardized Principle Component', *Photogrammetric Engineer. Remote Sens.* **59**, 991–996.

Goodman, S. J., McCaul, E. W. Jr., Buechler, D. E., Knupp, K., and Driscoll, K.: 2000, 'The 1997–98 El Niño Event and Related Wintertime Lightning Variations in the Southeastern United States', *Geophys. Res. Lett.* **27**, 541–544.

Goward, S. N., Tucker, C. J., and Dye, D. G.: 1985, 'North American Vegetation Patterns Observed with the NOAA-7 Advanced Very High-Resolution Radiometer', *Vegetatio*, **64**, 3–14.

Griffith, D. A.: 1996, 'Introduction: The Need for Spatial Statistics', in Arlinghaus, S. L., Griffith, D. A., Arlinghaus, W. C., Drake, W. D. and Nystuen J. D. (eds.), *Practical Handbook of Spatial Statistics*, CRC Press, Boca Raton, pp. 1–15.

Gutman, G., Csiszar, I., and Romanov, P.: 1999, 'Using NOAA/AVHRR Products to Monitor EL Niño Impacts: Focus on Indonesia in 1997–98', *Bull. American Meteorol. Soc.* **81**, 1189–1205.

Hansen, J. W., Hodges, A. W., and Jones, J. W.: 1998, 'ENSO Influence on Agriculture in the Southeastern United States', *J. Climate* **11**, 404–411.

Holben, B. N.: 1986, 'Characteristics of Maximum-Value Composite Images from Temporal AVHRR Data', *Int. J. Remote Sensing* **7**, 1417–1434.

Holben, B. N., Kaufman, Y. J., and Kendall, J. D.: 1990, 'NOAA-11 AVHRR Visible and Near-IR Inflight Calibration', *Int. J. Remote Sensing* **11**, 1511–1519.

Justice, C. O., Townshend, J. R. G., Holben, B. N., and Tucker, C. J.: 1985, 'Analysis of the Phenology of Global Vegetation Using Meteorological Satellite Data', *Int. J. Remote Sensing* **6**, 1271–1318.

Katz, R. W., Parlange, M. B., and Tebaldi, C.: 2003, 'Stochastic Modeling of the Effects of Large-Scale circulation on Daily Weather in the Southeastern U.S.', *Clim. Change* **60**, 189–216.

Kiladis, G. N. and Diaz, H. F.: 1989, 'Global Climate Extremes Associated with Extremes of the Southern Oscillation', *J. Climate*, **2**, 1069–1090.

Kogan, F. N.: 2000, 'Satellite-Observed Sensitivity of World Land Ecosystems to El Niño/La Niña', *Rem. Sens. Env.* **74**, 445–462.

Kuehl, R. O.: 2000, *Design of Experiments: Statistical Principles of Research Design and Analysis*, 2nd edn., Duxbury/Thomson Learning, Pacific Grove, CA, 666 pp.

Legendre, P.: 1993, 'Spatial Autocorrelation: Trouble or New Paradigm?', *Ecology* **74**, 1659–1673.

Li, Z. and Kafatos, M.: 2000, 'Interannual Variability of Vegetation in the United States and its Relation to El Niño/Southern Oscillation', *Rem. Sens. Env.* **71**, 239–247.

Littell, R. C., Milliken, G. A., Stroup, W. W., and Wolfinger, R. D.: 1996, *SAS System for Mixed Models*, SAS Institute Inc, Cary, NC, 633 pp.

Liu, W. T. and Negrón Juárez, R. I.: 2001, 'ENSO Drought Onset Prediction in Northeast Brazil Using NDVI', *Int. J. Remote Sensing* **17**, 3483–3501.

Loveland, T. R., Merchant, J. W., Ohlen, D. O., and Brown, J. F.: 1991, 'Development of a Land-Cover Characteristics Database for the Conterminous United States', *Photogrammetric Engineer. Remote Sens.* **57**, 1453–1463.

McCabe G. J. and Denttinger, M. D.: 1999, 'Decadal Variations in the Strength of ENSO Teleconnections with Precipitation in the Western United States', *Int. J. Clim.* **19**, 1399–1410.

Mennis J.: 2001, 'Exploring Relationships between ENSO and Vegetation Vigor in the South-East U.S.A. Using AVHRR Data', *Int. J. Remote Sensing* **22**, 3077–3092.

Moulin, S. A., Bondeau, A., and Delecolle, R.: 1998, 'Combining Agricultural Crop Models and Satellite Observations: From Field to Regional Scales', *Int. J. Remote Sensing* **19**, 1021–1036.

NOAA/USDA, Joint Agricultural Weather Facility: 1989–2000, *Weekly Weather and Crop Bulletin*, Vols. 77–87, NOAA/NWS/NCEP/CPC, Camp Springs, Maryland, published weekly.

Rajagopalan, B., Cook, E., U. Lall, and Ray, B. K.: 2000, 'Spatiotemporal Variability of ENSO and SST Teleconnections to Summer Drought over the United States during the Twentieth Century', *J. Climate* **13**, 4244–4255.

Ropelewski, C. F. and Halpert, M. S.: 1986, 'North American Precipitation and Temperature Patterns Associated with EL Niño/Southern Oscillation (ENSO)', *Mon. Wea. Rev.* **114**, 2352–2362.

Sellers, P. J., 1985: 'Canopy Reflectance, Photosynthesis and Transpiration', *Int. J. Remote Sensing* **6**, 1335–1372.

Trenberth, K. E.: 1997, 'The Definition of El Niño', *Bull. Amer. Meteorol. Soc.* **78**, 2771–2777.

Trenberth, K. E. and Hoar T.: 1996, 'The 1990–1995 El Niño-Southern Oscillation Event: Longest on Record', *Geophys. Res. Lett.* **23**, 57–60.

Tucker, C. J. Dregne, H. E., and Newcomb, W. W.: 1991, 'Expansion and Contraction of the Sahara Desert from 1980 to 1990', *Science* **253**, 299–301.

Tucker, C. J. and Sellers, P. J.: 1986, 'Satellite Remote Sensing of Primary Production', *Int. J. Remote Sensing* **7**, 1396–1416.

Unganai, L. S. and Kogan, F. N.: 1998, 'Drought Monitoring and Corn Yield Estimation in Southern Africa from AVHRR Data', *Rem. Sens. Env.* **63**, 219–232.

(Received 29 May 2002; in revised form 31 March 2003)

STOCHASTIC MODELING OF THE EFFECTS OF LARGE-SCALE CIRCULATION ON DAILY WEATHER IN THE SOUTHEASTERN U.S.

RICHARD W. KATZ [1], MARC B. PARLANGE [2] and CLAUDIA TEBALDI [1,3]

[1] *Environmental and Societal Impacts Group, National Center for Atmospheric Research, 1850 Table Mesa Drive, P.O. Box 3000, Boulder, CO 80307-3000, U.S.A.*
E-mail: rwk@ucar.edu

[2] *Department of Geography and Environmental Engineering, Johns Hopkins University, Baltimore, MD 21218, U.S.A.*

[3] *Research Applications Program, National Center for Atmospheric Research, Boulder, CO 80307, U.S.A.*

Abstract. Statistical methodology is devised to model time series of daily weather at individual locations in the southeastern U.S. conditional on patterns in large-scale atmosphere–ocean circulation. In this way, weather information on an appropriate temporal and spatial scale for input to crop–climate models can be generated, consistent with the relationship between circulation and temporally and/or spatially aggregated climate data (an exercise sometimes termed 'downscaling'). The Bermuda High, a subtropical Atlantic circulation feature, is found to have the strongest contemporaneous correlation with seasonal mean temperature and total precipitation in the Southeast (in particular, stronger than for the El Niño–Southern Oscillation phenomenon). Stochastic models for time series of daily minimum and maximum temperature and precipitation amount are fitted conditional on an index indicating the average position of the Bermuda High. For precipitation, a multi-site approach involving a statistical technique known as 'borrowing strength' is applied, constraining the relationship between daily precipitation and the Bermuda High index to be spatially the same. In winter (the time of greatest correlation), higher daily maximum and minimum temperature means and higher daily probability of occurrence of precipitation are found when there is an easterly shift in the average position of the Bermuda High. Methods for determining aggregative properties of these stochastic models for daily weather (e.g., variance and spatial correlation of seasonal total precipitation) are also described, so that their performance in representing low frequency variations can be readily evaluated.

1. Introduction

Many apparent connections between patterns in large-scale atmosphere–ocean circulation and climate in the southeastern U.S. have been noted in the literature. Prominent enough to be cited in studies ranging from national to global in scope is a tendency for winters to have below average temperature and above average precipitation during El Niño–Southern Oscillation (ENSO) warm phase events (Kiladis and Diaz, 1989; Ropelewski and Halpert, 1986). The climate in the Southeast has also been linked to the hemispheric-scale circulation, including the Pacific/North American (PNA) pattern (Leathers et al., 1991) and the North Atlantic Oscillation

Climatic Change **60:** 189–216, 2003.

(NAO) (Rogers, 1984), and to features of the Atlantic subtropical circulation such as the Bermuda High (BH) (Stahle and Cleaveland, 1992).

One reason for focusing on the Southeast is that effects of large-scale circulation, particularly the ENSO phenomenon, have been detected in crop yields; for instance, corn yields tend to be higher during ENSO cold phase events (Hansen et al., 1998, 1999). These impacts on crop yields have also been quantified indirectly through use of mechanistic crop models such as EPIC (Erosion Productivity Impact Calculator; Williams et al., 1989) (Izaurralde et al., 1999; Legler et al., 1999). In related work, Peters et al. (2003) found that a vegetation index, derived from remotely sensed data for the Southeast, has a negative response to ENSO warm phase events. Crop yield can be viewed as a complex integrator of weather over time. In fact, typical crop–climate models require weather inputs on a daily time scale, including minimum and maximum temperature and precipitation amount. Streamflow, another integrator of weather (especially precipitation) over both space and time, tends to be enhanced in the Gulf of Mexico region during ENSO warm phase events (Kahya and Dracup, 1993).

The purpose of the present paper is to statistically model the relationship between daily weather at individual locations in the Southeast and large-scale atmosphere–ocean circulation. The methodology involves fitting stochastic weather generators (e.g., Parlange and Katz, 2000; Richardson, 1981; Wilks and Wilby, 1999) conditional on circulation indices (Katz and Parlange, 1993, 1996; Woolhiser et al., 1993). Because of difficulties in detecting the weaker circulation signal in precipitation, the statistical technique known as 'borrowing strength' is applied (Tukey, 1961). This multi-site approach involves constraining the effect of the circulation index on a daily weather statistic to be the same at all locations within a specified region.

Among the indices of the BH, NAO, ENSO, and PNA averaged seasonally, the one having the strongest contemporaneous correlation with seasonal total precipitation and mean minimum and maximum temperature at eight locations in the Southeast is identified. The statistical methodology is then applied to model the relationship between the selected index and time series of daily minimum and maximum temperature and precipitation amount at the same locations. Through this statistical approach, weather data on an appropriate temporal and spatial scale for input to crop–climate models can be produced, consistent with the relationship between the circulation index and seasonally aggregated climate data (an exercise sometimes termed 'downscaling'; Wilby and Wigley, 1997). Methods for determining aggregative properties of these stochastic models for daily weather (e.g., variance and spatial correlation of seasonal total precipitation as aggregated from daily weather generated by these models) are also described.

Table I

List of acronyms for atmosphere–ocean circulation features and indices

Feature	Acronym
Bermuda pressure	BER
Bermuda High	BH
Bermuda High index	BHI
El Niño–Southern Oscillation	ENSO
North Atlantic Oscillation	NAO
New Orleans pressure	NO
New Orleans pressure index	NOI
Pacific/North Atlantic pattern	PNA
Southern Oscillation	SO

2. Background

Several studies have specifically focused on connections between temporally and/or spatially aggregated climate in the Southeast and large-scale atmosphere–ocean circulation patterns (Table I lists acronyms for circulation features). Upon closer examination, the link between ENSO and temperature and precipitation mentioned in the Introduction actually appears to be somewhat limited in extent both temporally (i.e., only during winter season) and spatially (i.e., only for part of Florida, as well as other portions of the Gulf Coast). In a systematic study of the relationship between large-scale circulation and precipitation in the Southeast, Henderson and Vega (1996) found a statistically significant correlation with ENSO only in Florida and only during the winter. In their study of precipitation during the spring in the southeastern coastal states of Georgia, North Carolina, and South Carolina, Stahle and Cleaveland (1992) did not detect any statistically significant correlation with ENSO. Neither did Downton and Miller (1993) for winter mean (or minimum) temperature averaged over central Florida.

Warmer than normal winters tend to occur in the Southeast with strong zonal flow over the North Atlantic, a possible connection to the NAO (Rogers, 1984). For instance, Downton and Miller (1993) obtained statistically significant correlations between winter mean (or minimum) temperature averaged over central Florida and the NAO. But no connection to precipitation in the Southeast has been identified, with Henderson and Vega (1996) finding essentially no correlation. Likewise, Stahle and Cleaveland (1992) did not detect any correlation between the NAO and spring precipitation in the southeastern coastal states (not including Florida).

Lower (higher) than normal temperatures over the Southeast are associated with a meridional (zonal)-like flow pattern over the continental U.S., particularly in winter, a possible link to the PNA (Leathers et al., 1991). For example, Downton and Miller (1993) obtained statistically significant correlations between winter mean (or minimum) temperature averaged over central Florida and the PNA. For precipitation, any relationship to PNA is anticipated to be weaker and primarily in winter, with meridional flow being associated with lower than normal precipitation inland (i.e., west of the Appalachian Mountains), but not necessarily in coastal areas (i.e., east of the Appalachians) where the relationship might even be reversed (Leathers et al., 1991). In particular, Henderson and Vega (1996) obtained a statistically significant negative correlation between the PNA and winter precipitation for the inland ('North Central') portion of the Southeast, but a positive correlation for Florida, and otherwise no relationship along the East Coast. Stahle and Cleaveland (1992) could not find any connection between the PNA and spring precipitation for the southeastern coastal states (not including Florida).

In attempts to identify sources of low frequency variations in the climate of the Southeast, features of the Atlantic subtropical circulation have recently received some consideration as well. One such feature is the BH or North Atlantic subtropical anticyclone (Davis et al., 1997). Stahle and Cleaveland (1992) provided evidence that spring precipitation in the southeastern coastal states (not including Florida) is lower (higher) than normal if the western edge of the subtropical high is west (east) of its average position. Henderson and Vega (1996) found more robust correlations between seasonal total precipitation and the BH than with ENSO, NAO, or PNA, the relationship always being statistically significant irrespective of season or subregion of the Southeast. Connections to the temperature in the Southeast have received less attention, but Henderson and Muller (1997) did suggest that temperature extremes in the adjacent South Central region of the U.S. are related to the BH.

Although much research has been devoted to the relationship between large-scale atmosphere–ocean circulation and spatial and/or temporally aggregated climate, not nearly as much work has been conducted on the corresponding relationships to daily weather data at individual sites, whether for the Southeast or elsewhere. Usually, the approach has been to let the data 'speak for themselves' (e.g., Legler et al., 1999). That is, the daily weather observations are 'composited' according to a categorization of circulation patterns (e.g., El Niño, neutral, and La Niña events), without an explicit examination of the circulation effects on these scales. The limitations of this approach relate to the non-uniqueness in disaggregation (or downscaling). For instance, does a decrease in monthly or seasonal precipitation correspond to fewer storms, less intense storms, or both?

A few studies have directly examined the relationship between circulation and daily weather statistics. For example, Wettstein and Mearns (2002) found relationships between the NAO and daily minimum and maximum temperature statistics in the northeastern U.S. and Canada. Through resampling, Thompson and Wallace (2001) established statistically significant links between the NAO and daily weather statistics at locations across the Northern Hemisphere. A more systematic approach consists of fitting a stochastic model to time series of daily weather conditional on the state of a circulation index. This approach was followed by Katz and Parlange (1993) in modeling daily precipitation at individual locations in California conditional on a pressure index and by Woolhiser et al. (1993) in modeling daily precipitation at individual locations in the southwestern U.S. conditional on ENSO. Subsequent applications include Grondona et al. (2000) to model the relationship between ENSO and daily precipitation at stations in Argentina and Uruguay and Kiely et al. (1998) to model daily precipitation at a site in Ireland conditional on pressure or wind direction.

3. Data and Seasonal Statistics

3.1. DATA

Figure 1 shows the eight locations across the Southeast for which climate data were obtained, with the station names (and abbreviations) and period of record being listed in Table II. It should be noted that only one of these locations is in Florida or near the Gulf Coast. All these sites are first-order stations (as classified by the U.S. National Weather Service), with the advantage of high quality measurements (in particular, not much missing data; see Table II for percentage missing data), but the disadvantage of possible urban influences (e.g., heat island effect). The daily weather variables considered are minimum and maximum temperature (°C) and precipitation amount (mm), with a 'trace' (i.e., a positive amount smaller than 0.005 in ($\approx$0.13 mm)) being treated as a zero amount. Seasonal means and totals are obtained by aggregating the daily data over the appropriate time period, with the seasonal statistic being computed if three or less days of observations within the season are missing (otherwise the seasonal value is treated as missing).

To reflect low frequency variation, each circulation measure is aggregated to a single seasonal value. The source of the data for the BH circulation measure is the gridded monthly mean sea level pressure from the NCAR/NCEP reanalysis (http://www.cgd.ucar.edu/cas/catalog/nmc/rean/press/means.html; Kalnay et al., 1996). The approach of Stahle and Cleaveland (1992) was followed, with an exploratory analysis in which the pressure at each of the individual grid points in a region that spans the eastern U.S., Gulf of Mexico, and western Atlantic Ocean was considered as a candidate. Two grid points were selected: one at 30° N, 90° W (near New Orleans), the second at 32.5° N, 65° W (near Bermuda) (denoted by NO and BER, respectively; see Figure 1). The New Orleans grid point is the same as, and the one for Bermuda slightly different from, those selected by Stahle and Cleaveland (1992).

A seasonal measure of the BH is constructed by first aggregating the monthly mean pressure to seasonal means and then taking the difference (Bermuda minus New Orleans). This measure effectively indicates the approximate position of the

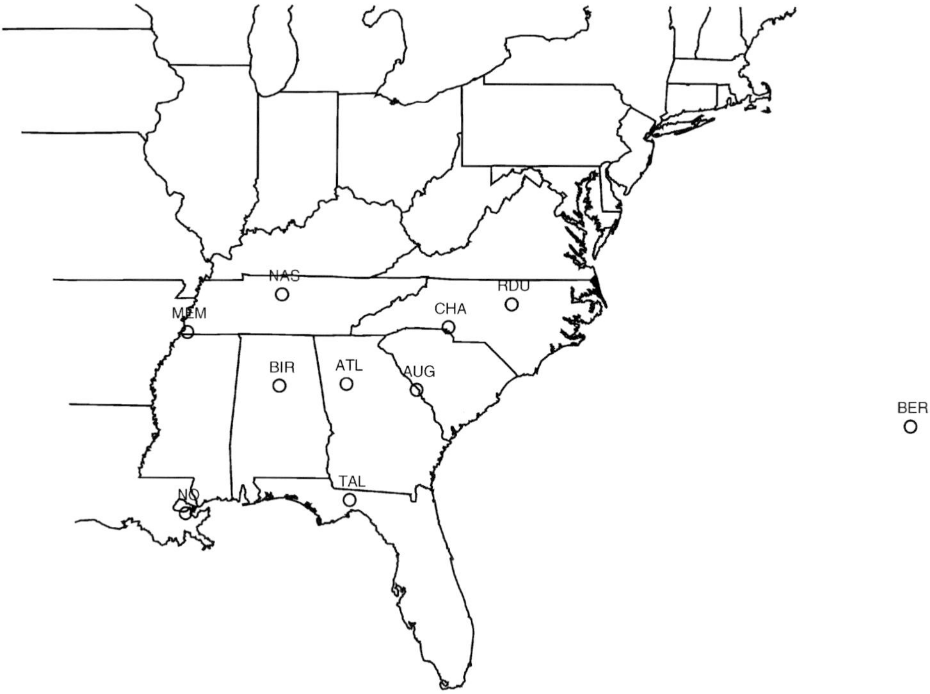

Figure 1. Map of southeastern U.S. showing eight sites of daily weather data, as well as two pressure grid locations on which BH based.

Table II
List of abbreviations for stations and characteristics of climate data record for winter season (e.g., winter 1959 = December 1958–February 1959)

Site	Abbrev.	Time period	Missing data (%)		
			Min. temp.	Max. temp.	Prec.
Atlanta, GA	ATL	1959–1996	0	0	0
Augusta, GA	AUG	1959–1996	0	0	0
Birmingham, AL	BIR	1959–1996	1.8	1.8	2.8
Charlotte, NC	CHA	1959–1996	0	0	0
Memphis, TN	MEM	1959–1996	0	0	0
Nashville, TN	NAS	1959–1996	0.8	0.8	0.8
Raleigh–Durham, NC	RDU	1959–1995	0	0	0
Tallahassee, FL	TAL	1959–1995	0	0	0.1

western edge of the Atlantic subtropical high, with higher values indicating a location farther east than normal. In most of the subsequent analyses, the BH will be converted into a simpler two-state index denoted by BHI (i.e., BHI = 1 if BH > median, BHI = 0 otherwise). The pressure at New Orleans alone (NO) will also be used as a circulation measure, somewhat cruder but still related to the position of the subtropical high (in this case, higher values indicate a location farther west than normal). The corresponding two-state index is denoted by NOI (i.e., NOI = 1 if NO > median, NOI = 0 otherwise).

Three other circulation measures are considered: (i) an index of the NAO based on the difference in standardized seasonal mean sea level pressure between Ponta Delgada, Azores and Stykkisholmur/Reykjavik, Iceland (source: http://www.cgd.ucar.edu/~jhurrell/nao.html; Hurrell, 1995); (ii) an index of the PNA derived from the standardized 500 hPa geopotential heights at four grid points (i.e., 20° N, 160° W; 45° N, 165° W; 55° N, 115° W; 30° N, 85° W) across the Pacific Ocean and North America, including one in northern Florida (source: http://tao.atmos.washington.edu/data_sets/pna/; Wallace and Gutzler, 1981); and (iii) an index of the Southern Oscillation (SO) based on the difference in standardized seasonal mean sea level pressure between Tahiti and Darwin (source: http://www.cpc.ncep.noaa.gov/data/indices/index.html; Trenberth, 1984).

3.2. SEASONAL STATISTICS

Tables III–V list the sample correlations between the circulation indices and winter mean minimum and maximum temperature and total precipitation at the southeastern sites. These correlations are presented only for the winter season, the time of the year in which they are greatest (although spring and summer seasons would be more important for most crops). To check for time trends, the variable ‘Year’ is included in the tables as well.

The sample correlations of winter temperature with the BH are all statistically significant at the 5% level, being of a similar magnitude for both the mean minimum and maximum (Tables III and IV). But for the NO, all of the correlations with the maximum and all but two with the minimum are statistically insignificant. In all cases for the NAO and at all but one location for the PNA, the correlations with both the minimum and maximum are statistically significant, but they are systematically smaller in magnitude than the corresponding ones for the BH. The SO performs worst of all the circulation indices for the minimum temperature, with all correlations being statistically insignificant, and second worst to the NO for the maximum, with all but one being insignificant. Linear time trends for minimum temperature are statistically significant at several locations, most likely associated with the urban heat island. Nevertheless, these temperature–circulation correlations do not change much if a linear trend is removed from the temperature time series. For this reason, the temperature time series are not de-trended in subsequent analyses.

Table III

Correlation between circulation indices (as well as time trend) and winter mean minimum temperature at eight locations in Southeast (a correlation of about 0.32 or greater in absolute value is statistically significant at the 5% level)

Site	BH	NO	NAO	PNA	SO	Year
ATL	0.661	–0.179	0.604	–0.340	–0.042	0.432
AUG	0.785	–0.314	0.455	–0.492	0.031	0.248
BIR	0.795	–0.380	0.461	–0.586	0.167	0.157
CHA	0.717	–0.212	0.564	–0.426	–0.033	0.458
MEM	0.573	–0.299	0.485	–0.276	–0.082	0.425
NAS	0.713	–0.384	0.525	–0.362	–0.093	0.253
RDU	0.749	–0.290	0.510	–0.473	0.040	0.395
TAL	0.726	–0.293	0.382	–0.500	0.032	0.093

Table IV

Same as Table III except for winter mean maximum temperature

Site	BH	NO	NAO	PNA	SO	Year
ATL	0.636	–0.038	0.633	–0.395	0.171	0.396
AUG	0.689	–0.053	0.474	–0.517	0.242	0.255
BIR	0.734	–0.203	0.505	–0.510	0.262	0.235
CHA	0.651	0.022	0.567	–0.369	0.156	0.245
MEM	0.593	–0.174	0.473	–0.316	0.093	0.292
NAS	0.713	–0.181	0.572	–0.390	0.089	0.166
RDU	0.727	–0.110	0.531	–0.494	0.185	0.272
TAL	0.736	–0.181	0.530	–0.646	0.335	0.296

For winter total precipitation (Table V), the sample correlations with the BH are not as great as for temperature, but still statistically significant for the 'inland' half of the stations (i.e., Atlanta, Birmingham, Memphis, and Nashville). If the NO pressure alone is used instead, then the correlations with the remaining 'coastal' half of the stations (i.e., Augusta, Charlotte, Raleigh–Durham, and Tallahassee) are at least borderline statistically significant. The pattern in the precipitation correlations with the PNA resembles that for the BH (if the sign is reversed), most being smaller in magnitude but still attaining at least borderline statistical significance at the inland locations. The precipitation correlations with the NAO and

Table V

Same as Table III except for winter total precipitation

Site	BH	NO	NAO	PNA	SO	Year
ATL	0.522	–0.422	0.185	–0.277	0.111	–0.148
AUG	0.235	–0.480	–0.008	0.107	–0.181	–0.156
BIR	0.401	–0.406	0.028	–0.387	0.239	–0.210
CHA	0.341	–0.385	0.086	–0.183	0.021	–0.162
MEM	0.493	–0.088	0.277	–0.419	0.036	0.185
NAS	0.375	–0.166	0.121	–0.430	0.109	–0.055
RDU	0.260	–0.310	0.056	–0.040	–0.029	–0.039
TAL	0.098	–0.280	–0.282	0.023	–0.241	–0.133

SO are all statistically insignificant, although a few for the NAO attain borderline significance. There is no evidence of a trend in precipitation.

To examine the conditional distribution of seasonally aggregated climate, the two-state BHI and NOI are used. Figures 2–5 show boxplots of the conditional distribution of minimum and maximum winter mean temperature given the BHI state (for ease of display the stations are divided into the inland and coastal groups). Although the conditional median is higher at all eight locations given BHI = 1, a consistent effect is not evident for variability, as measured in terms of the interquartile range (i.e., height of box) (plus the shifts in the conditional standard deviation are not always in the same direction as for the interquartile range). For winter total precipitation, both the conditional median and interquartile range are higher at the four inland locations given BHI = 1 (Figure 6); likewise, at the four coastal locations given NOI = 0 (Figure 7). In summary, warmer and wetter than normal conditions on the average (along with higher precipitation variability) during the winter in the Southeast tend to be associated with the western edge of the Atlantic subtropical high being east of its average position. The statistical significance of these apparent effects will be evaluated more systematically in Sections 4 and 5.

Table VI lists the sample correlations between the different atmospheric–ocean circulation indices in winter. The BH has statistically significant correlations with the NAO and PNA, but not with the SO. The correlation between the pressure at the two grid points, BER and NO, on which the BH is based is not statistically significant, with the BH itself being more strongly related to the BER than to the NO. The only statistically significant trend is for the NAO, whose apparent non-stationarity has been noted in the literature (Hurrell, 1995).

Making use of multiple regression analysis, the inclusion of the SO (or other circulation variables) as an additional predictor, along with the BH, generally does not result in a statistically significant improvement in the fit to seasonal temperature or precipitation. Likewise, replacing the BH with the two gridded pressure

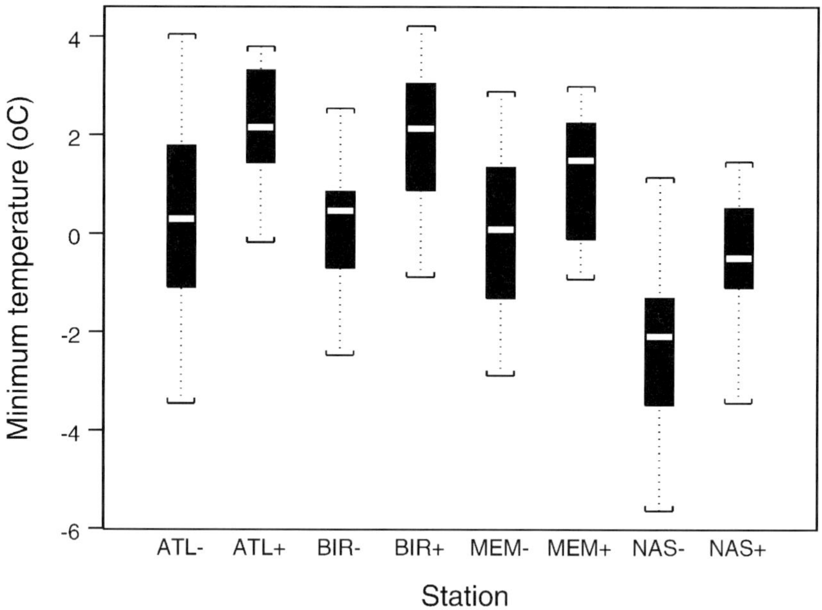

Figure 2. Boxplots (i.e., sample minimum, lower quartile, median, upper quartile, and maximum) of conditional distribution of winter mean minimum temperature at four inland locations given state of BHI (BHI = 1 indicated by '+' after station abbreviation, BHI = 0 by '–').

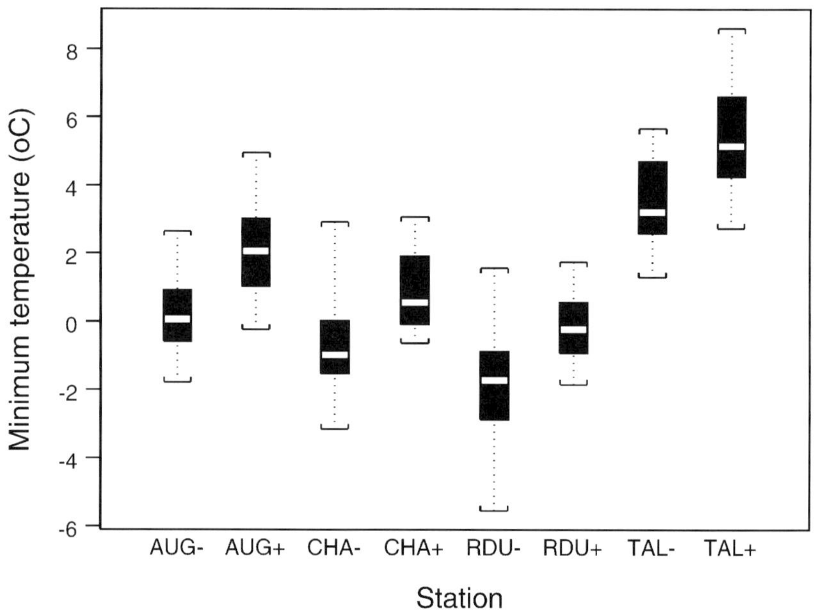

Figure 3. Same as Figure 2 except at four coastal locations.

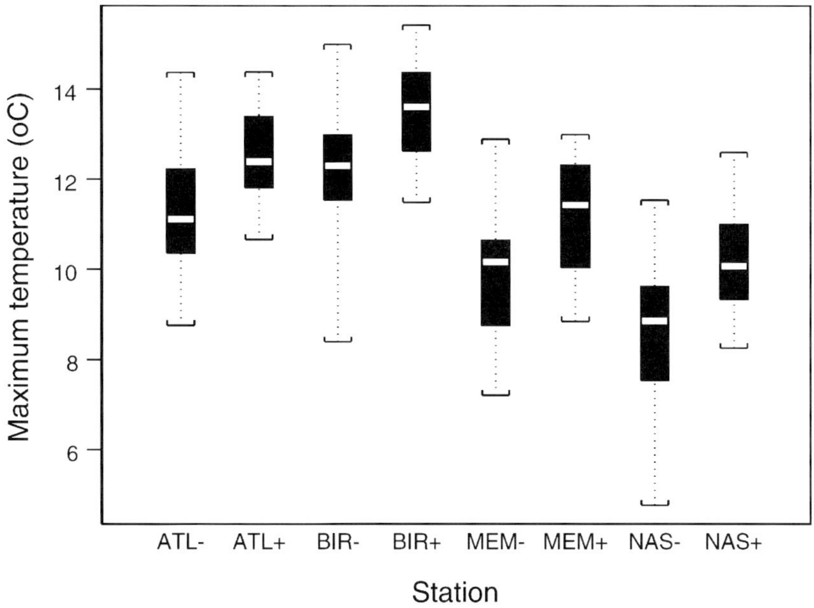

Figure 4. Boxplots of conditional distribution of winter mean maximum temperature at four inland locations given state of BHI.

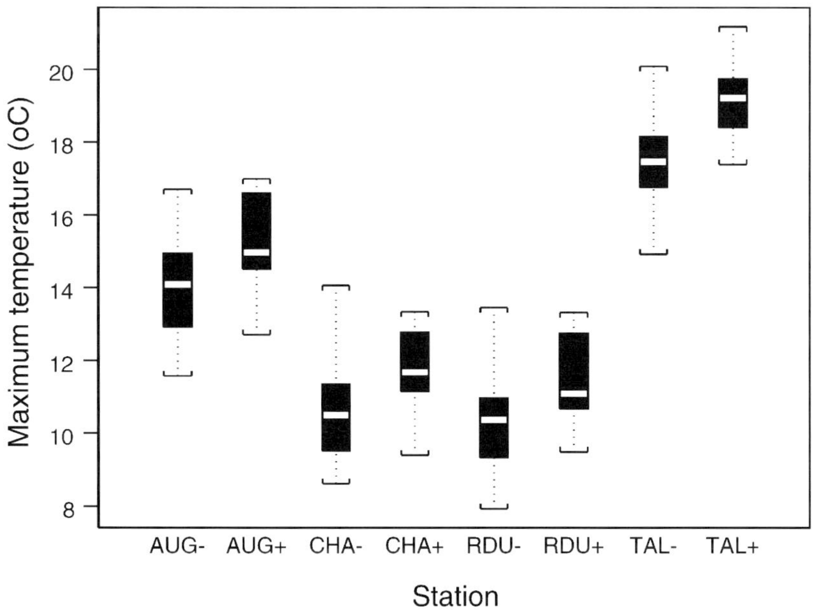

Figure 5. Same as Figure 4 except at four coastal locations.

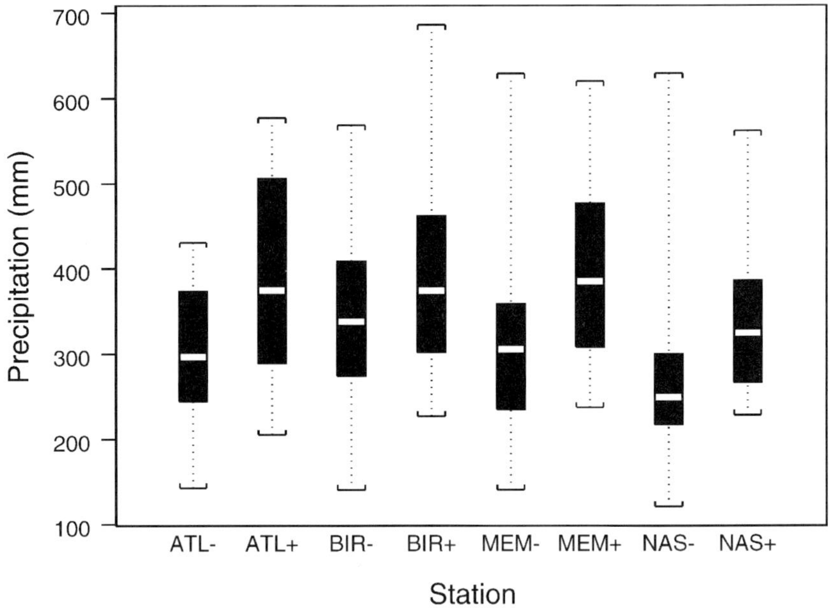

Figure 6. Boxplots of conditional distribution of winter total precipitation at four inland locations given state of BHI.

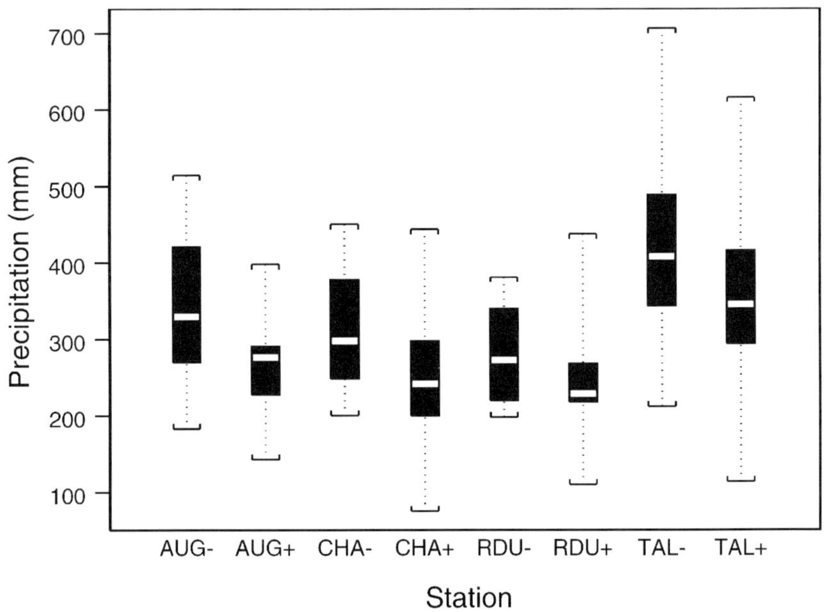

Figure 7. Same as Figure 6 except at four coastal locations given state of NOI.

Table VI

Correlations between circulation indices in winter (a correlation of about 0.32 or greater in absolute value is statistically significant at the 5% level)

Index	Year	BER	BH	NO	NAO	PNA
BER	0.166	–	–	–	–	–
BH	0.040	0.781	–	–	–	–
NO	0.177	0.221	–0.437	–	–	–
NAO	0.427	0.656	0.475	0.203	–	–
PNA	0.103	–0.504	–0.565	0.157	0.011	–
SO	–0.156	0.309	0.220	0.102	0.051	–0.402

variables, BER and NO (or introducing BER into the regression along with the NO in the case of seasonal precipitation for the coastal sites), usually produces only marginal improvements in the fit. Consequently, the subsequent analyses of time series of daily weather are restricted to conditioning on only the BHI or NOI.

4. Daily Temperature: Single-Site Approach

Stochastic weather generators typically model the minimum and maximum temperature at the same site as a bivariate time series, incorporating both the cross correlation on the same day and the serial correlation from one day to the next. Usually, dependence between temperature and precipitation is introduced by shifting the temperature means and standard deviations conditional on precipitation occurrence (e.g., Richardson, 1981). In the present paper, to simplify matters the daily minimum and maximum temperature are analyzed separately, with no dependence on precipitation occurrence. The temperature is treated at one location at a time, conditioning on the two-state BHI and focusing on the winter season.

4.1. METHODS

Let $\{X_t, t = 1, 2, \ldots, T\}$ denote the time series of daily temperature (either minimum or maximum) during a given season of length T days in a particular year at a single location. This time series is assumed stationary (i.e., the annual cycle is assumed small enough to ignore within a season). The temporal dependence structure of daily temperature is modeled as a first-order autoregressive [AR(1)] process, but with the parameters varying with the BHI state. Given BHI $= i$, the daily temperature time series is assumed to be an AR(1) process with conditional mean

μ_i, standard deviation σ_i, and first-order autocorrelation coefficient ϕ_i, $i = 0, 1$. That is, the temperature today X_t is related to the temperature yesterday X_{t-1} by

$$X_t = \mu_i + \phi_i(X_{t-1} - \mu_i) + \sigma_a(i)a_t\,, \quad a_t \sim N(0, 1). \tag{1}$$

Here the error component a_t is serially uncorrelated and has a standard normal distribution (denoted by $N(0, 1)$ in (1)). The standard deviation σ_i of the temperature variable is related to the standard deviation of the error component $\sigma_a(i)$ of the AR(1) process by

$$\sigma_i^2 = \sigma_a^2(i)/(1 - \phi_i^2) \tag{2}$$

(e.g., Mearns et al., 1984). The assumption that, conditional on BHI, the daily temperature time series can be modeled as an AR(1) process is not as restrictive as it might appear. Unconditionally, such an approach induces a model for daily temperature that is more complex than AR(1) (Katz and Parlange, 1996).

Special cases of the dependence of these AR(1) parameters on the BHI can be fitted by this modeling approach, ranging from 'no effect' to 'full effect', by imposing the following constraints in (1):

(i) *no effect* $[\mu_0 = \mu_1, \sigma_a(0) = \sigma_a(1), \phi_0 = \phi_1]$
(ii) *mean effect* $[\sigma_a(0) = \sigma_a(1), \phi_0 = \phi_1]$
(iii) *mean and variance effect* $(\phi_0 = \phi_1)$
(iv) *full effect* (no constraints).

A description of the methodology involved in fitting such models appears in Appendix A. In essence, the approach uses multiple regression analysis with least squares parameter estimation to fit (1) under the constraints imposed by models (i)–(iv). In this way, the autocorrelation of the temperature time series is explicitly taken into account.

Which of these alternative forms of model best fits the data is decided on the basis of the Bayesian information criterion (BIC) (see Appendix A). This technique balances the improved fit of more complex models against the need to estimate additional parameters (Schwarz, 1978). According to this criterion, the model with the smallest BIC value is preferred.

4.2. RESULTS

The assumption that the minimum and maximum temperature time series are stationary within the winter is indirectly examined through the comparison of boxplots for the distribution of the corresponding monthly means. Figure 8 shows these boxplots for the months of December, January, and February at Atlanta. On average, January is the coolest of the three months, February the warmest. No pattern in variability, as measured by the interquartile range, is evident. Similar patterns in monthly mean minimum and maximum temperature during the winter are evident in the boxplots (not shown) for the other locations. By allowing some

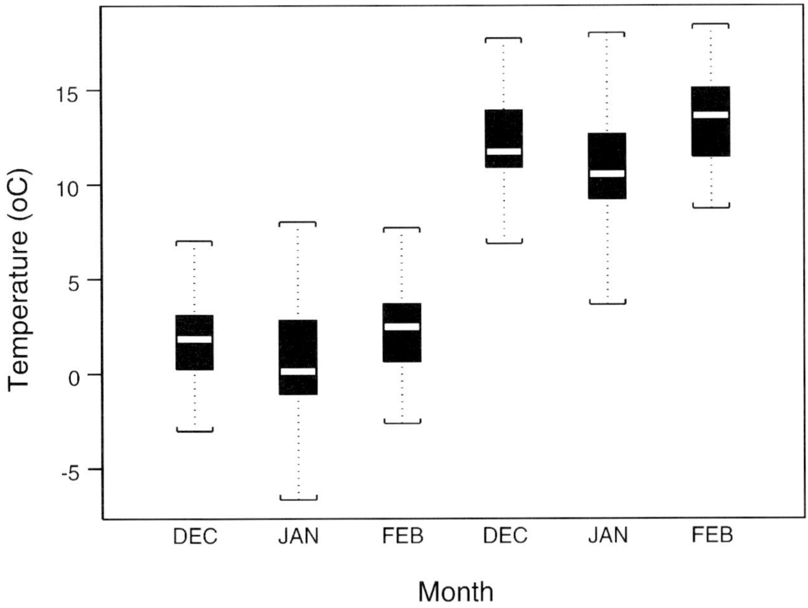

Figure 8. Boxplots of distribution of monthly mean minimum and maximum temperature during winter at Atlanta.

of the parameters in (1) to depend on the time of year, it would be straightforward to introduce annual cycles into the methodology described in Section 4.1.

Tables VII and VIII list the sample conditional mean, standard deviation, and first-order autocorrelation coefficient for the daily time series of minimum and maximum temperature in winter at the eight locations in the Southeast. For the minimum temperature (Table VII), the conditional mean is 1–2 °C higher when BHI = 1, whereas the shifts in conditional standard deviation and autocorrelation are relatively small and inconsistent in direction. For the maximum temperature (Table VIII), the increases in conditional mean when BHI = 1 generally are only slightly smaller than the corresponding ones for the minimum temperature. The daily time series of minimum and maximum temperature possess substantial sample autocorrelations, usually between 0.6 and 0.7.

Tables VII and VIII include a summary of the results of applying the model fitting and selection procedure to the winter daily time series of minimum and maximum temperature at the eight locations in the Southeast. With only two exceptions, the BIC selects model (ii) (i.e., varying the mean with the BHI state) as best. For the maximum temperature at Charlotte and Raleigh–Durham, model (iii) (i.e., varying both the mean and variance) is selected instead. In no case is model (i) (i.e., no effect) or model (iv) (i.e., varying the mean, variance, and autocorrelation) selected.

How the model selection criterion performs in practice is demonstrated for the example of the daily maximum temperature in winter at Atlanta. The BIC values

5. Daily Precipitation: Multi-Site Approach

The analysis of the effects of circulation on daily precipitation is complicated by several factors. For one thing, the intermittency of the precipitation process necessitates separate treatment of two component processes, the occurrence and intensity. There is no way of anticipating whether effects of the BH (or NO) on seasonal total precipitation correspond to effects on the daily occurrence alone, the daily intensity alone, or on both. For another, the correlations between the BH (or NO) and seasonal total precipitation are weaker than the corresponding ones for temperature. As will be seen, a single-site analysis of precipitation is not sufficiently powerful to obtain statistical significance when examining these effects on a daily time scale.

5.1. METHODS

Let $\{J_t, t = 1, 2, \ldots, T\}$ denote the time series of daily precipitation occurrence (i.e., $J_t = 1$ indicates a 'wet' day, $J_t = 0$ 'dry' day). This time series is assumed stationary within the winter season. To model the tendency of wet or dry spells to persist, a two-state, first-order Markov chain is assumed as a model for $\{J_t\}$, but with the parameters varying with the BHI or NOI (e.g., Katz and Parlange, 1993). Conditional on circulation index state i, the time series of daily precipitation occurrence is assumed to be a first-order Markov chain with transition probabilities

$$P_{jk}(i) = \Pr\{J_t = k \mid J_{t-1} = j\}, \quad j, k = 0, 1. \tag{3}$$

Let the probability of a wet day be denoted by $\pi_i = \Pr\{J_t = 1\}$ and the first-order autocorrelation coefficient (or 'persistence' parameter) by $d_i = \mathrm{corr}(J_{t-1}, J_t)$, likewise conditional on index state i. These quantities are related to the transition probabilities by

$$\pi_i = P_{01}(i)/(1 - d_i), \quad d_i = P_{11}(i) - P_{01}(i) \tag{4}$$

(e.g., Katz, 1996). The two transition probabilities that appear in (4), $P_{01}(i)$ and $P_{11}(i)$, are sufficient to specify the probabilistic properties of the Markov chain. The assumption that the daily precipitation occurrence time series, given the circulation index, is a first-order Markov chain is not as restrictive as it might appear. Unconditionally, this approach induces a model for daily precipitation occurrence that is more complex than a first-order Markov chain (Katz and Parlange, 1996).

Let $P_{jk}(i, s)$ denote the transition probability defined in (3) at site s, $s = 1, 2, \ldots, S$, where S denotes the total number of sites in the region. The *borrowing strength* approach constrains the effect of the circulation index on the transition probability $P_{jk}(i, s)$ to be the same for all sites within a given region. Specifically, it is assumed that the ratio of the transition probability under the two index states,

$$\Delta_{jk}(s) = P_{jk}(1, s)/P_{jk}(0, s), \tag{5}$$

Table X

Conditional probability of occurrence, transition probabilities, and median intensity for winter daily precipitation at four inland locations given state of BHI

Site	Prob. of occurrence		Transition prob. (P_{01})		Transition prob. (P_{11})		Median intensity (mm)	
	BHI = 0	BHI = 1	BHI = 0	BHI = 1	BHI = 0	BHI = 1	BHI = 0	BHI = 1
ATL	0.324	0.377	0.252	0.295	0.479	0.520	5.59	5.84
BIR	0.312	0.374	0.241	0.293	0.468	0.514	6.86	6.35
MEM	0.296	0.346	0.233	0.278	0.451	0.482	5.59	5.84
NAS	0.327	0.373	0.265	0.296	0.457	0.507	4.83	5.33

is constant, say $\Delta_{jk}(s) = \Delta_{jk}$, for all locations s, $s = 1, 2, \ldots, S$. Such a constraint might well be a plausible approximation in the present circumstances, in which the indices are based on large-scale circulation features. The assumption (5) imposes no constraint on how the precipitation climatology varies across sites (e.g., probability of wet day), only on its relationship with the circulation index. The advantage of this approach is that fewer parameters need to be estimated (e.g., with four locations, conditioning on a two-state index introduces only one additional parameter instead of four for a given transition probability), so that there is a higher chance of detecting an effect that is stable across space.

Details on how to estimate the transition probabilities of the Markov chain model for precipitation occurrence under the constraint (5) are discussed in Appendix B. The BIC is again used to select the appropriate form of conditioning on circulation (see Appendix B). The key requirement is that, conditional on the state of the circulation index, daily precipitation occurrence be spatially independent. The hope is that, if the conditioning variable has a strong enough relationship with precipitation occurrence, then sufficient unconditional dependence would be induced (e.g., Hughes et al., 1999). This assumption will be scrutinized further in Section 6. Instead of (5), an alternative approach would be to impose the constraint in terms of the logistic transformation of precipitation probability (Katz and Parlange, 1995; Stern and Coe, 1984; Woolhiser et al., 1993).

5.2. RESULTS

The assumption that the precipitation occurrence time series is stationary is indirectly examined through the comparison of boxplots for the distribution of the corresponding monthly precipitation totals. Figure 10 shows these boxplots for the months of December, January, and February at Atlanta. No consistent pattern in either the median or the interquartile range is evident, nor in the boxplots (not shown) for the other locations.

Table X includes sample estimates of the conditional probability of occurrence of precipitation and intensity (i.e., precipitation amount on wet days) median in

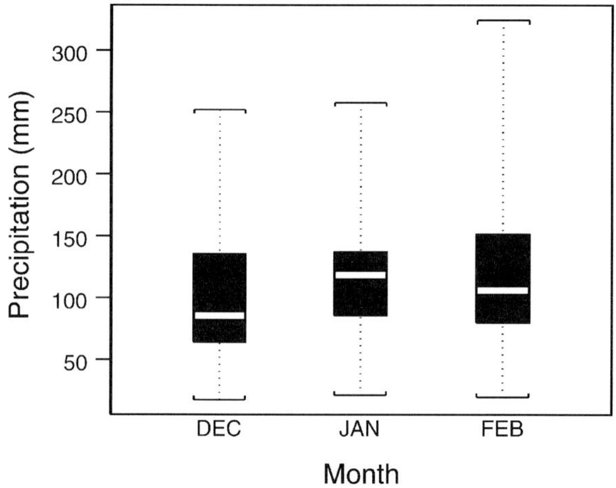

Figure 10. Same as Figure 8 except for total precipitation at Atlanta.

Table XI

Same as Table X except at four coastal locations given state of NOI

Site	Prob. of occurrence		Transition prob. (P_{01})		Transition prob. (P_{11})		Median intensity (mm)	
	NOI = 0	NOI = 1	NOI = 0	NOI = 1	NOI = 0	NOI = 1	NOI = 0	NOI = 1
AUG	0.342	0.299	0.272	0.233	0.481	0.459	6.10	5.08
CHA	0.352	0.301	0.273	0.248	0.497	0.430	5.33	4.83
RDU	0.331	0.304	0.267	0.243	0.461	0.445	5.84	5.08
TAL	0.317	0.273	0.259	0.228	0.444	0.392	8.13	6.60

winter at the four inland locations given the state of the BHI. The estimated probability of a wet day ranges from 14–20% higher when BHI = 1. On the other hand, the shift in the sample median precipitation intensity with the BHI is relatively small and inconsistent in direction. Table XI includes the same statistics at the four coastal locations given the state of the NOI. Consistent shifts in the probability of precipitation are again obtained, being 9–17% higher when NOI = 0. The median intensity is always higher when NOI = 0, with these shifts still being relatively small but somewhat larger in magnitude than those associated with the BHI at the inland locations.

Table X also provides the estimated transition probabilities at the inland locations given the BHI state, with P_{01} and P_{11} being 12–21% and 7–11% higher, respectively, when BHI = 1. Table XI gives the same statistics at the coastal locations conditional on the NOI, with P_{01} and P_{11} being 10–17% and 4–16% higher, respectively, when NOI = 0. Using same methodology as in Katz and Parlange (1993) applied to a single site at a time (i.e., maximum likelihood estimates of transition probabilities and BIC for model selection), these shifts in transition

Table XII

Model selection for winter daily precipitation occurrence at inland locations conditional on BHI ($m = 38$ yr, $S = 4$ sites)

Model	Number parameters	Proportionate effect Δ_{01}	Δ_{11}	BIC
No effect	8	1	1	16,676.5
Borrow strength (P_{01})	9	1.171	1	16,661.3
Borrow strength (P_{11})	9	1	1.091	16,673.3
Borrow strength (P_{01}, P_{11})	10	1.171	1.091	16,658.1 [a]
Full effect	16	[1.115, 1.214]	[1.070, 1.111]	16,687.3

[a] Minimum.

probabilities were not found statistically significant at either the inland locations conditional on the BHI or the coastal locations conditional on the NOI.

For the borrowing strength approach, the BIC values are included in Table XII for the daily precipitation occurrence transition probabilities at the four inland locations conditional on the BHI. The best model involves varying both of the transition probabilities, P_{01} and P_{11}, conditional on the BHI, but constraining their effect to be same at each location (i.e., borrowing strength in both cases). The second-best model involves borrowing strength for P_{01}, but with P_{11} not depending on the BHI. It is noteworthy that the alternative model in which both transition probabilities at all four locations vary with the BHI, but the borrowing strength constraint is not imposed (i.e., consistent with a single-site approach), is inferior to no conditioning on the BHI. The same ordering of BIC values is obtained for the precipitation occurrence transition probabilities at the four coastal locations conditional on the NOI (not shown).

These effects on the transition probabilities (e.g., Δ_{01} is estimated as about 1.17 and Δ_{11} as 1.09 for the inland locations conditional on BHI; 0.89 and 0.92, respectively, for the coastal locations conditional on NOI) can be converted via (4) into the corresponding effects on the probability of occurrence and persistence parameter. In this case, they correspond primarily to shifts in π (smoothed versions of the unconstrained estimates listed in Tables X and XI, not d). An analogous borrowing strength approach could be applied to precipitation intensity, but the effects of BHI and NOI are too small to attain statistical significance.

6. Aggregative Properties of Stochastic Weather Models

An AR(1) model, as in (1), can be used to generate daily time series of minimum or maximum temperature and these daily values then aggregated to seasonal means. For this reason, the aggregative properties of such models are of interest. The daily mean parameter in (1) is equivalent to the seasonal mean. However, the variance of the seasonal mean temperature depends not only on the daily variance, but also on the first-order autocorrelation coefficient (e.g., chapter 17 in von Storch and Zwiers, 1999). The shift in the mean of daily minimum and maximum temperature, but not necessarily in the variance or autocorrelation, with the BHI (Tables VII and VIII) is consistent with the corresponding result that only the median (or mean) of seasonal mean temperatures is apparently related to the BHI (Figures 2–5).

Likewise, the aggregative properties of stochastic models for time series of daily precipitation amount are of interest. It is convenient to assume a 'chain-dependent process' as a stochastic model for daily precipitation amount, a model that includes two components: (i) a first-order Markov chain model for occurrence as in (3); and (ii) intensities that are independent and identically distributed (Katz and Parlange, 1993; Stern and Coe, 1984). The mean of seasonal total precipitation is directly proportional to both the probability of occurrence of precipitation and the mean intensity (see Appendix C). Yet the seasonal variance is not only directly proportional to the intensity variance, but also nonlinearly related to the probability of precipitation, the persistence parameter of the Markov chain, and the mean intensity (see Appendix C). In particular, an increase in just the probability of a wet day with the BHI or NOI (i.e., as supported by the results in Tables X and XI) will produce an increase in both the mean and variance of seasonal total precipitation as in Figures 6 and 7 (Katz, 1993).

Previous studies of the aggregative properties of stochastic weather models have primarily focused on the tendency for daily stochastic models to underestimate the variance of monthly or seasonal total precipitation (e.g., Mavromatis and Hansen, 2001). This underestimation has sometimes been attributed to the neglect of low frequency sources of variability (Hansen and Mavromatis, 2001), with one proposed remedy being to condition on large-scale circulation measures (Katz and Parlange, 1993, 1996).

Table XIII lists the seasonal standard deviation of winter total precipitation at the four inland locations in the Southeast, as 'induced' by a chain-dependent process for daily precipitation amount with different degrees of conditioning on the BHI. This conditioning includes: (i) no dependence of parameters (denoted by 'none' in the table); (ii) only the transition probabilities of the Markov chain model for occurrence vary according to the constraint (5) (denoted by 'borrow'); and (iii) unconstrained variation of the transition probabilities, as well as of the intensity mean and variance (denoted by 'full'). Appendix C sketches how these induced standard deviations can be obtained (also see Katz and Parlange, 1993, 1996). Unlike the situation in some other regions (e.g., U.S. West Coast in winter; Katz and Parlange, 1993), the chain-dependent process with no conditioning does not underestimate the observed standard deviation (also listed in the table) by very much, except perhaps at Nashville. Moreover, neither type of conditioning increases the standard deviation much relative to that for no conditioning. This behavior is attributable to the relatively weak effect of the BHI on precipitation in

Table XIII

Standard deviation (mm) of winter total precipitation at four inland locations, as induced by conditional stochastic models for daily precipitation given BHI

Site	Conditioning			Observed
	None	Borrow	Full	
ATL	99.5	103.1	109.2	103.9
BIR	108.7	112.3	111.3	111.9
MEM	109.5	112.7	117.2	115.7
NAS	87.9	90.9	96.0	103.1

Table XIV

Spatial correlation of winter total precipitation between pairs of the four inland locations, as induced by conditional stochastic models for daily precipitation given BHI (observed correlation in parentheses)

Site	ATL		BIR		MEM	
BIR	0.09	(0.74)	–	–	–	–
MEM	0.15	(0.28)	0.07	(0.35)	–	–
NAS	0.15	(0.32)	0.07	(0.43)	0.12	(0.77)

the Southeast, perhaps reflecting a more general lack of evidence of any substantial source of low frequency variation.

The spatial correlations between winter total precipitation at the inland locations, as induced by the stochastic models for daily precipitation, can also be examined. An explanation of how to obtain these induced spatial correlations is provided in Appendix C (also see Katz, 2002). Table XIV lists the induced spatial correlations for pairs of the four inland locations in the case of full conditioning (i.e., not imposing the constraints of the borrowing strength approach) of the parameters of the chain-dependent process on the BHI state, along with the observed correlation values. Despite the relatively weak connection with the BH, this approach does produce some positive spatial correlations, but the values are much smaller than those observed. These results cast some doubts on the assumption of conditional spatial independence of precipitation occurrence given the BHI state, made in conjunction with the borrowing strength approach (Section 5).

7. Discussion

This paper deals with the development of a statistical methodology to model time series of daily weather at individual locations conditional on patterns in large-scale atmosphere–ocean circulation. Exploiting the stability of the relationship between circulation and a daily weather statistic across a region, the methodology includes a multi-site approach based on the statistical technique of borrowing strength. The approach is applied to model daily time series of daily minimum and maximum temperature and precipitation amount at locations in the Southeastern U.S. conditional on indices of the Bermuda High. It is established that in winter the daily mean minimum and maximum temperature and probability of precipitation occurrence all tend to be higher when the position of the Bermuda High is farther east than average. No relationship with other daily weather statistics, such as the minimum or maximum temperature variance or the median precipitation intensity, is identified.

Such results may appear disappointing, perhaps even in conflict with previously published results on the relationship between circulation indices and spatially and temporally aggregated climate data. Yet they most likely reflect the difficulties that arise in disaggregation (or downscaling), particularly for variables like precipitation. In particular, determining the extent to which a correlation with seasonal total precipitation corresponds to effects on storm frequency or intensity is a challenging statistical problem.

The methodological approach presented here could be extended in several respects. In the borrowing strength technique, the assumption of conditional spatial independence for precipitation occurrence could be relaxed and spatial dependence explicitly modeled. Rather than restricting attention to indices that assume only a small number of states, it would be straightforward to condition on a continuous index (Woolhiser et al., 1993). Finally, an alternative approach that should result in much stronger relationships involves conditioning on high, instead of low, frequency information about circulation (i.e., smaller-scale circulation indices that vary on a daily time scale) (Bellone et al., 2000; Tebaldi, 2000).

Acknowledgements

We thank Linda Mearns for encouragement to work on this topic, Larry McDaniel for acquiring daily weather data, and Ruth Doherty for comments. Two anonymous reviewers are acknowledged for their careful reading of the manuscript. Research supported in part by EPA/NCERQA grant R824997-01-C and NASA/MTPE grant OA99073, W-19, 080, as well as NSF grants DMS-9312686 and DMS-9815344 to the Geophysical Statistics Project of the National Center for Atmospheric Research. NCAR is sponsored by the National Science Foundation.

Appendix A: Model Fitting for Daily Temperature

The stochastic model for time series of daily temperature (1) can be reexpressed as:

$$X_t = \beta_0 + \beta_1 X_{t-1} + \beta_2 i + \beta_3 i X_{t-1} + \epsilon_t(i), \ \ \mathrm{var}[\epsilon_t(i)] = \sigma_\epsilon^2(i). \tag{A1}$$

Here β_0, β_1, β_2, and β_3 are regression parameters, $\sigma_\varepsilon(i)$ the standard deviation of the error term (i.e., equivalent to $\sigma_a(i)$ in (1)), and i the circulation index state (i.e., 0 or 1). Models (i)–(iv) listed in Section 4.1 can all be fitted by multiple regression analysis, imposing constraints in (A1): (i) $\beta_2 = \beta_3 = 0$, $\sigma_\varepsilon(0) = \sigma_\varepsilon(1)$; (ii) $\beta_3 = 0$, $\sigma_\varepsilon(0) = \sigma_\varepsilon(1)$; (iii) $\beta_3 = 0$; and (iv) no constraints.

For a regression model of the form (A1), the BIC selects the model for which the following quantity is a minimum (e.g., Chapter 6 in Venables and Ripley, 1999):

$$\mathrm{BIC}(p) = n_0 \ln s_\epsilon^2(0) + n_1 \ln s_\epsilon^2(1) + p \ln m. \tag{A2}$$

Here $s_\varepsilon(i)$ denotes the root mean square error for the fitted model on the subset of the data for which the index $= i$ (consisting of a total of n_i days). The first two terms on the right-hand side of (A2) constitute the goodness-of-fit component, whereas the last term is a penalty for the number of parameters p estimated and depends on the sample size of m years as well (for a discussion of how to choose the appropriate sample size, see Kass and Wasserman, 1995).

Appendix B: Borrowing Strength for Daily Precipitation Occurrence

Let $n_{jk}(i, s)$ denote the 'transition count' (i.e., number of times precipitation occurrence state j is followed by state k) given circulation index state i at site s. The common effect Δ_{jk} is estimated from $n_{jk}(i, s)$ by

$$\hat{\Delta}_{jk} = [n_{jk}(1,\cdot)/n_{j\cdot}(1,\cdot)]/[n_{jk}(0,\cdot)/n_{j\cdot}(0,\cdot)]. \tag{B1}$$

Here $n_{jk}(i,\cdot) = \Sigma_s n_{jk}(i, s)$, $n_{j\cdot}(i,\cdot) = \Sigma_s n_{j\cdot}(i, s)$ and $n_{j\cdot}(i, s) = n_{j0}(i, s) + n_{j1}(i, s)$.

Then $P_{jk}(i, s)$ can be obtained from Δ_{jk} and $P_{jk}(s)$ (i.e., the transition probability without conditioning on the index) through the relationship

$$P_{jk}(1, s) = [\Delta_{jk} P_{jk}(s)]/[1 - w_j(s) + \Delta_{jk} w_j(s)] \tag{B2}$$

and (5). Here $w_j(s) = [w\ n_{j\cdot}(1, s)]/[(1 - w)n_{j\cdot}(0, s) + w\ n_{j\cdot}(1, s)]$, where w denotes the fraction of years for which the index assumes state 1.

The BIC selects the Markov chain model for which the following quantity is a minimum (e.g., Katz and Parlange, 1993):

$$\mathrm{BIC}(p) = -2\Sigma_{s,i,j,k}\, n_{jk}(i, s) \ln \hat{P}_{jk}(i, s) + p \ln(mS). \tag{B3}$$

Here p denotes number of parameters a given model requires be estimated and m the number of years of data at the S sites. In (B3), $\hat{P}_{jk}(i, s)$ is the estimate obtained by first estimating Δ_{jk} through (B1) and then substituting this estimate into (B2) and (5) ($P_{jk}(s)$ in (B2) is estimated by conventional maximum likelihood (e.g., Katz and Parlange, 1993).

Appendix C: Induced Variance and Correlation for Seasonal Total Precipitation

Let S_T denote the daily precipitation amount totaled over T days; that is, $S_T = Z_1 + Z_2 + \cdots + Z_{N(T)}$, where Z_l denotes the precipitation intensity on the lth wet day and $N(T)$ the total number of wet days. The conditional mean and variance of S_T given a circulation index state i can be expressed by (e.g., Katz and Parlange, 1993):

$$\begin{aligned} E(S_T \mid i) &= T\pi_i\mu_Z(i), \\ \mathrm{var}(S_T \mid i) &\approx T\{\pi_i\sigma_Z^2(i) + \pi_i(1-\pi_i)[(1+d_i)]/(1-d_i)]\mu_Z^2(i)\}, \\ &\quad \text{for large } T. \end{aligned} \tag{C1}$$

Here $\mu_Z(i)$ and $\sigma_Z^2(i)$ denote the mean and variance of daily intensity given index state i.

The unconditional (or 'induced') variance is related to the conditional means and variances (C1) by (Katz and Parlange, 1996):

$$\begin{aligned} \mathrm{var}(S_T) &= (1-w)\mathrm{var}(S_T \mid 0) + w\,\mathrm{var}(S_T \mid 1) \\ &\quad + w(1-w)[E(S_T \mid 1) - E(S_T \mid 0)]^2. \end{aligned} \tag{C2}$$

Let the total precipitation at two sites be denoted by S_T and S'_T, respectively. Under the assumption that S_T and S'_T are conditionally independent given the index, the induced correlation between S_T and S'_T is related to the conditional means and variances (C1) by (Katz, 2002):

$$\begin{aligned} \mathrm{corr}(S_T, S'_T) &= \{w(1-w)[E(S_T \mid 1) - E(S_T \mid 0)][E(S'_T \mid 1) \\ &\quad - E(S'_T \mid 0)]\}/[\mathrm{var}(S_T)\mathrm{var}(S'_T)]^{1/2}. \end{aligned} \tag{C3}$$

References

Bellone, E., Hughes, J. P., and Guttorp, P.: 2000, 'A Hidden Markov Model for Downscaling Synoptic Atmospheric Patterns to Precipitation Amounts', *Clim. Res.* **15**, 1–12.

Davis, R. E., Hayden, B. P., Gay, D. A., Phillips, W. L., and Jones, G. V.: 1997, 'The North Atlantic Subtropical Anticyclone', *J. Climate* **10**, 728–744.

Downton, M. W. and Miller, K. A.: 1993, 'The Freeze Risk to Florida Citrus. Part II: Temperature Variability and Circulation Patterns', *J. Climate* **6**, 364–372.

Grondona, M. O., Podestá, G. P., Bidegain, M., Marino, M., and Hordij, H.: 2000, 'A Stochastic Precipitation Generator Conditioned on ENSO Phase: A Case Study in Southeastern South America', *J. Climate* **13**, 2973–2986.

Hansen, J. W., Hodges, A. W., and Jones, J. W.: 1998, 'ENSO Influences on Agriculture in the Southeastern United States', *J. Climate* **11**, 404–411.

Hansen, J. W., Jones, J. W., Kiker, C. F., and Hodges, A. W.: 1999, 'El Niño-Southern Oscillation Impacts on Winter Vegetable Production in Florida', *J. Climate* **12**, 92–102.

Hansen, J. W. and Mavromatis, T.: 2001, 'Correcting Low-Frequency Variability Bias in Stochastic Weather Generators', *Agric. For. Meteorol.* **109**, 297–310.

Henderson, K. G. and Muller, R. A.: 1997, 'Extreme Temperature Days in the South-Central United States', *Clim. Res.* **8**, 151–162.

Henderson, K. G. and Vega, A. J.: 1996, 'Regional Precipitation Variability in the Southern United States', *Phys. Geog.* **17**, 93–112.

Hughes, J. P., Guttorp, P., and Charles, S. P.: 1999, 'A Non-Homogeneous Hidden Markov Model for Precipitation Occurrence', *Applied Statistics* **48**, 15–30.

Hurrell, J. W.: 1995, 'Decadal Trends in the North Atlantic Oscillation: Regional Temperatures and Precipitation', *Science* **269**, 676–679.

Izaurralde, R. C., Rosenberg, N. J., Brown, R. A., Legler, D. M., Tiscareño López, M., and Srinivasan, R.: 1999, 'Modeled Effects of Moderate and Strong "Los Niños" on Crop Productivity in North America', *Agric. For. Meteorol.* **94**, 259–268.

Kahya, E. and Dracup, J. A.: 1993, 'U.S. Streamflow Patterns in Relation to the El Niño/Southern Oscillation', *Water Resour. Res.* **29**, 2491–2503.

Kalnay, E. et al.: 1996, 'The NCEP/NCAR 40-Year Reanalysis Project', *Bull. Amer. Meteorol. Soc.* **77**, 437–472.

Kass, R. E. and Wasserman, L.: 1995, 'A Reference Bayesian Test for Nested Hypotheses and its Relationship to the Schwarz Criterion', *J. American Statistical Association* **90**, 928–934.

Katz, R. W.: 1993, 'Towards a Statistical Paradigm for Climate Change', *Clim. Res.* **2**, 167–175.

Katz, R. W.: 1996, 'Use of Conditional Stochastic Models to Generate Climate Change Scenarios', *Clim. Change* **32**, 237–255.

Katz, R. W.: 2002, 'Techniques for Estimating Uncertainty in Climate Change Scenarios and Impact Studies', *Clim. Res.* **20**, 167–185.

Katz, R. W. and Parlange, M. B.: 1993, 'Effects of an Index of Atmospheric Circulation on Stochastic Properties of Precipitation', *Water Resour. Res.* **29**, 2335–2344.

Katz, R. W. and Parlange, M. B.: 1995, 'Generalizations of Chain-Dependent Processes: Application to Hourly Precipitation', *Water Resour. Res.* **31**, 1331–1341.

Katz, R. W. and Parlange, M. B.: 1996, 'Mixtures of Stochastic Processes: Application to Statistical Downscaling', *Clim. Res.* **7**, 185–193.

Kiely, G., Albertson, J. D., Parlange, M. B., and Katz, R. W.: 1998, 'Conditioning Stochastic Properties of Daily Precipitation on Indices of Atmospheric Circulation', *Meteorol. Appl.* **5**, 75–87.

Kiladis, G. N. and Diaz, H. F.: 1989, 'Global Climate Anomalies Associated with Extremes in the Southern Oscillation', *J. Climate* **2**, 1069–1090.

Leathers, D. J., Yarnal, B., and Palecki, M. A.: 1991, 'The Pacific/North American Teleconnection Pattern and United States Climate. Part I: Regional Temperature and Precipitation Associations', *J. Climate* **4**, 517–528.

Legler, D. M., Bryant, K. J., and O'Brien, J. J.: 1999, 'Impact of ENSO-Related Climate Anomalies on Crop Yields in the U.S.', *Clim. Change* **42**, 351–375.

Mavromatis, T. and Hansen, J. W.: 2001, 'Interannual Variability Characteristics and Simulated Crop Response of Four Stochastic Weather Generators', *Agric. For. Meteorol.* **109**, 283–296.

Mearns, L. O., Katz, R. W., and Schneider, S. H.: 1984, 'Extreme High-Temperature Events: Changes in their Probabilities with Changes in Mean Temperature', *J. Clim. Appl. Meteorol.* **23**, 1601–1613.

Parlange, M. B. and Katz, R. W.: 2000, 'An Extended Version of the Richardson Model for Simulating Daily Weather Variables', *J. Appl. Meteorol.* **39**, 610–622.

Peters, A. J., Ji, L., and Walter-Shea, E.: 2003, 'Southeastern U.S. Vegetation Response to ENSO Events (1989–1999)', *Clim. Change* **60**, 175–188.

Richardson, C. W.: 1981, 'Stochastic Simulation of Daily Precipitation, Temperature, and Solar Radiation', *Water Resour. Res.* **17**, 182–190.

Rogers, J. C.: 1984, 'The Association between the North Atlantic Oscillation and the Southern Oscillation in the Northern Hemisphere', *Mon. Wea. Rev.* **112**, 1999–2015.

Ropelewski, C. F. and Halpert, M. S.: 1986, 'North American Precipitation and Temperature Patterns Associated with the El Niño/Southern Oscillation (ENSO)', *Mon. Wea. Rev.* **114**, 2352–2362.

Schwarz, G.: 1978, 'Estimating the Dimension of a Model', *Ann. Statistics* **6**, 461–464.

Stahle, D. W. and Cleaveland, M. K.: 1992, 'Reconstruction and Analysis of Spring Rainfall over the Southeastern U.S. for the Past 1000 Years', *Bull. Amer. Meteorol. Soc.* **73**, 1947–1961.

Stern, R. D. and Coe, R.: 1984, 'A Model Fitting Analysis of Daily Rainfall Data (with discussion)', *J. Royal Statistical Society, Series A* **147**, 1–34.

Tebaldi, C.: 2000, 'Influence of Large Scale Circulation Measures on Precipitation at Local Stations in the South East of the U.S.', NRCSE-TRS No. 55, Univ. of Washington, Seattle (available at *http://www.nrcse.washington.edu/research/reports.html*).

Thompson, D. W. J. and Wallace, J. M.: 2001, 'Regional Climate Impacts of the Northern Hemisphere Annular Mode', *Science* **293**, 85–89.

Trenberth, K. E.: 1984, 'Signal versus Noise in the Southern Oscillation', *Mon. Wea. Rev.* **112**, 326–332.

Tukey, J. W.: 1961, 'Data Analysis and Behavioral Science or Learning to Bear the Quantitative Man's Burden by Shunning Badmandments', Previously Unpublished Manuscript Reprinted in Jones, L. V. (ed.), 1986, *The Collected Works of John W. Tukey, V. III: Philosophy and Principles of Data Analysis: 1949–1964*, Wadsworth, Monterey, CA.

Venables, W. N. and Ripley, B. D.: 1999, *Modern Applied Statistics with S-PLUS*, 3rd edn., Springer, New York.

von Storch, H. and Zwiers, F. W.: 1999, *Statistical Analysis in Climate Research*, Cambridge University Press, Cambridge.

Wallace, J. M. and Gutzler, D. S.: 1981, 'Teleconnections in the Geopotential Height Field during the Northern Hemisphere Winter', *Mon. Wea. Rev.* **109**, 784–812.

Wettstein, J. J. and Mearns, L. O.: 2002, 'The Influence of the North Atlantic–Arctic Oscillation on Mean, Variance, and Extremes of Temperature in the Northeastern United States and Canada', *J. Climate* **15**, 3586–3600.

Wilby, R. L. and Wigley, T. M. L.: 1997, 'Downscaling General Circulation Model Output: A Review of Methods and Limitations', *Prog. Phys. Geog.* **21**, 530–548.

Wilks, D. S. and Wilby, R. L.: 1999, 'The Weather Generation Game: A Review of Stochastic Weather Models', *Prog. Phys. Geog.* **23**, 329–357.

Williams, J. R., Jones, C. A., Kiniry, J. R., and Spaniel, D. A.: 1989, 'The EPIC Crop Growth Model', *Trans. Amer. Soc. Agric. Eng.* **32**, 497–511.

Woolhiser, D. A., Keefer, T. O., and Redmond, K. T.: 1993, 'Southern Oscillation Effects on Daily Precipitation in the Southwestern United States', *Water Resour. Res.* **29**, 1287–1295.

(Received 29 May 2002; in revised form 3 April 2003)